Hargrave Jennings

Nature worship

An account of phallic faiths and practices: Ancient and modern

Hargrave Jennings

Nature worship
An account of phallic faiths and practices: Ancient and modern

ISBN/EAN: 9783337282691

Printed in Europe, USA, Canada, Australia, Japan

Cover: Foto ©berggeist007 / pixelio.de

More available books at **www.hansebooks.com**

NATURE WORSHIP

AN ACCOUNT OF

Phallic Faiths & Practices

ANCIENT AND MODERN;

INCLUDING THE

ADORATION OF THE MALE AND FEMALE

POWERS IN VARIOUS NATIONS

AND THE

Sacti Puja of Indian Gnosticism

BY THE AUTHOR OF "PHALLICISM."

PRIVATELY PRINTED,

MDCCCXCI.

PREFACE.

THE exhaustion of the first edition of *Phallicism*, of which only five hundred copies were printed, and the continued demand for the same led to the consideration of the question whether we should reprint our first issue or prepare a new work embodying material not therewith incorporated. It having been resolved to adopt the latter course, with the exception of a few matters of definition, necessary in every edition, this volume will be found to present information not contained in *Phallicism* or in any of the supplementary volumes on the *Worship of the Serpent*, of *Trees*, of *Fishes*, *Flowers*, *Fire*, &c., afterwards issued.

The value of the following treatise lies in the fact that the information it imparts has not been derived from speculative or partisan ideas; neither is it the fruit of the inventive genius of the romance writer: it consists rather of matter gathered from an immense variety of Eastern writers, and in making it as full and explanatory as possible, the very best authorities have been pressed into the service, and the substance of their statements duly quoted.

Other books of a phallic character, so far as they have been of any use or value, have been made up exclusively of extracts from Indian and other works; their fault has been that they have failed to acknowledge the sources to which they were indebted for their information. The reader will find in our pages frequent reference to the authors cited, and an honest use of the marks which distinguish a quoted from an original passage.

CONTENTS.

Chapter I.—The Origin of Phallicism 1

Chapter II.—Deities connected with Phallic Worship......... 8

Chapter III.—Emblems connected with Phallic Worship 16

Chapter IV.—Phallic Objects 33

Chapter V.—Wide Prevalence of Phallic Worship 43

Chapter VI.—Phallic Worship in the Middle Ages 60

Chapter VII.—Moral Aspects of Phallicism 70

Chapter VIII.—Sacta Puja, the Worship of the Female Power 82

NATURE WORSHIP.

CHAPTER I.
The Origin of Phallicism.

BY way of definition of terms, very few words are needed in introducing our present treatise. Phallic worship (the phase of nature worship to which our attention will be directed) is the adoration of the human organs of generation, from φαλλός (membrum virile). Originally it was of a character free from all associations of licentiousness and indecency, — viewed of course in the light of remoter times—it simply represented allegorically, the mysterious union of male and female as the source of the continuation of the existence of animated beings. In course of time, however, it was attended with so much that was detrimental to the morals of the people, especially in the degenerate days of Rome, that the governing authorities had to interfere and in turn modify and suppress it.

Extraordinary as this worship, in its extensive and varied forms and wide-spread prevalance in almost every nation on the face of the earth, may well be termed, there will be found when it comes to be carefully regarded and studied, very much in it that for want of better information and education is both reasonable and natural. Men have always to some extent been observers of the processes constantly going on under the control of the ordinary laws of nature, and it was not likely that they would fail to be impressed with the profound mysteries connected with the propagation of their own and other species. "Two causes," says a writer on the subject, "must have forcibly struck the minds of men in early periods when observant of the operations of nature,—one the generative power, and the other the productive,—the active and passive causes. This double mode of production visible in nature, must have given rise to comparisons with the mode proceeding in the generation of animals, in which two causes concur, the one active and the other passive,—the one male, and the other female, the one as father, the other as mother. Nature to the early man was not brute matter, but a being invested with his own personality, and endowed with the same feelings and passions, and performing the same actions. Generation, begetting,—production, bringing forth, were thus his ideas of cause and effect. The earth was looked upon as the mould of nature, as the recipient of seeds, the nurse of what was produced in its bosom; the sky was the fecundating and fertilising power. An analogy was suggested in the union of the male and female."

Of the extensive prevalence of this worship we have ample evidence. It occurs in Egypt with the deity Khem, in India with Siva, in Assyria with Vul, in Greece with Pan and Priapus, in the Scandinavian and Tuetonic nations with Fricco, in Spain with Hortanes. It has been found in different parts of the American continent, in Mexico, Peru, and Hayti; in both these latter places numerous phalli modelled in clay have been discovered; and in the islands of the Pacific ocean, on festive occasions, a phallus highly ornamented, called by the natives Tinas, is carried in procession.

"Among the simple and primitive races of men, the act of generation was considered as no more than one of the operations of nature contributing to the reproduction of the species, as in agriculture the sowing of the seed for the production of corn, and was consequently looked upon as a solemn duty consecrated to the Deity; as Payne Knight remarks, 'it was considered as a solemn sacrament in honour of the Creator.'"

"The indecent ideas attached to the phallic symbol were, though it seems a paradox to say so, the result of a more advanced civilisation verging towards its decline, as we have evidence at Rome and Pompeii."

Voltaire says, speaking of the worship of Priapus:—"Our ideas of prosperity lead us to suppose that a ceremony which appears to us so infamous, could only be invented by licentiousness; but it is impossible to believe that depravity of manners would ever have led among any people to the establishment of religious ceremonies. It is probable, on the contrary, that this custom was first introduced in times of simplicity, that the first thought was to honour the deity in the symbol of life which it has given us. Such a ceremony may have excited licentiousness among youths, and have appeared ridiculous to men of education in more refined, more corrupt, and more enlightened times."

And Mr. Patterson, in the *Asiatic Researches*, some ninety years ago, wrote: "It is probable that the idea of obscenity was not originally attached to these symbols: and it is likely that the inventors themselves might not have foreseen the disorders which this worship would occasion amongst mankind. Profligacy eagerly embraces what flatters its propensities, and ignorance follows blindly, whenever example excites: it is therefore no wonder, that a general corruption of manners should ensue, increasing in proportion as the distance of time involved the original meaning of the symbol in darkness and oblivion. Obscene mirth became the principal feature of the popular superstition, and was, even in after times, extended to, and intermingled with, gloomy rites and bloody sacrifices."

The origin of Phallicism is involved in so much obscurity that nothing can be said thereon with any degree of certainty or clearness. The Phœnicians traced its introduction to Adonis, the Egyptians to Osiris, the Phrygians to Attys, the Greeks to Dionysus. The Scánda-purána and Visvasara-pracasa, or declaration of what is most excellent in the world, supplies us with the following legend, which seems to be nothing more or less than an illusion to the sun which in autumn loses its fructifying influence:

"One day Maha-Deva, the great Hindu god, was rambling over the earth, naked, and with a large club in his hand, he chanced to pass near the post where several Munis were performing their devotions. Maha-Deva laughed at them, insulted them in the most provoking and indecent terms, and lest his expressions should not be forcible enough, he accompanied the whole with significant signs and gestures. The offended Munis cursed him, and the linga or phallus fell to the ground. Maha-Deva, in this state of mutilation, travelled over the world, bewailing his misfortune. His consort, too, hearing of this accident, gave herself up to grief, and ran after him in a state of distraction, repeating mournful songs. This is what the Greek mythologists called the wanderings of Damater and the lamentations of Bacchus.

"The world being thus deprived of its vivifying principles, generation and vegetation were at a stand; gods and men were alarmed, but having discovered the cause of it, they all went in search of the sacred linga, and at last found it grown to an immense size, and endowed with life and motion.

"Having worshipped the sacred pledge, they cut it with their hatchets into one-and-thirty pieces, which, Polypus like, soon became perfect lingas. The Devatas left one-and-twenty of them upon earth, carried nine to heaven, and removed one into the inferior regions for the benefit of the inhabitants of the three worlds. One of these lingas was erected on the banks of the Camudvati, or Euphrates, under the name of Baleswaralinga, or the linga of Iswara, the infant, who seems to answer to the Jupiter Puer of the western mythologists.

"To satisfy Devi, and restore all things to their former situation, Maha-Deva was born again in the character of Baleswara, or Iswara, the infant. Baleswara, who fosters and preserves all, though a child, was of uncommon strength; he had a beautiful countenance; his manners were most engaging, and his only wish was to please everybody, in which he succeeded effectually; but his subjects waited with impatience till he came to the age of maturity, that he might bless them with an heir to his virtues. Baleswara, to please them, threw off his child-like appearance and suddenly became a man, under the title of Lileswara, or Iswara, who gives pleasure and delight. He then began to reign over gods and men, with the strictest observance of justice and equity; his subjects were happy, and the women beheld with delight his noble and manly appearance. With the view of doing good to mankind he put himself at the head of a powerful army, and conquered many distant countries, destroying the wicked and all oppressors; he had the happiness of his subjects so much at heart, that he entirely neglected any other pursuit. His indifference for the female sex alarmed his subjects; he endeavoured to please them, but his embraces were fruitless. This is termed *Aschalana* in Sanscrit, and the place where this happened was in consequence denominated *Aschalana-sthan*. The Apsaras, or celestial nymphs, tried in vain the effect of their charms. At last Samirama came to Aschalana-sthan, and retiring into a solitary place in its vicinity, chanted her own metamorphoses and those of Lileswara, who happening to pass by, was so delighted with the sweetness of her voice, that he went to her and enquired who she was. She related to him how they went together into Utcoladesa in the characters of the Capoteswara and Capotesi, adding, "You appeared then as Mocshewara, and I became Anaysa; you are now Lileswara and I am Samirama, but I shall be soon Lileswari." Lileswara being under the influence of Maya, or worldly illusion, did not recollect any of these transactions; but suspecting that the person he was speaking to might be a manifestation of Parvati, he thought it advisable to marry her; and having obtained her consent, he seized her hand, and led her to the performance of the nuptial ceremony, to the universal satisfaction of his subjects. Gods and men met to solemnise this happy union, and the celestial nymphs and heavenly quiristers graced it with their presence. Thus Samirama and Lileswara commenced their reign to the general satisfaction of mankind, who were happy under their virtuous administration.

"From that period the three worlds began to know and worship Lileswara, who, after he had conquered the universe returned into *Cusha-duipa*. Lileswara, having married Samirama, lived constantly with her, and followed her wherever she chose to go; in whatever pursuits and pleasures she delighted in these alone he took pleasure; thus they travelled over hills and through forests to distant countries; but at last returned to Cusha-duipa, and Samirama seeing a delightful grove near the Hradancita, (or deep water), with a small river of the same name, expressed a wish that he would fix the place of their residence in this beautiful spot, there to spend their days in pleasure."

The Hindus insist that the black stone in the wall of the Caaba is no other than the linga or phallus of Maha-Deva; and that, when the Caabs was rebuilt by Mohammed (as they appear to have been), it was placed in the wall out of contempt; but the new converted pilgrims would not give up the worship of the black stone; and sinistrous portents forced the ministers of the new religion to connive at it. Arabian authors also inform us that stones were worshipped all over Arabia, particularly at Mecca; and Alshahrestani says that the temple of Mecca was dedicated to Zohal or Kyevun, who is the same with Saturn.

The author of the Dabistan declares positively that the *Hejar-al-aswad*, or black stone, was the image of Kyevun. Though these accounts somewhat differ from those of the Puranas, yet they show that this black stone was the object of an idolatrous worship from the most remote times. *

Faber says: "It is possible that the phallus received its name from the Palli, Pelasgi, or Palestini,—the Pelasgi being much addicted to phallic worship. Miphlezeph, the idol of Maacha, seems to be Ma-Phallasath, the great phallic goddess. I think

---

* See *Asiatic Researches*, vol. iii.

that she was rather Venus or Mylitta, than Priapus, as Selden supposes. Perhaps, however, we ought to derive Palli, Pelasgi, Palestini, neither from Peleg nor Felasge; but to deduce these several appellatims from phallus, rather than what I just observed, *vice versa*. It certainly is not improbable, that those nations were so called from the worship of the phallus, since the Hindoos assure us, that the Ionians or Yonijas received their name from the devotion to the mysteries of the Yoni."

Mr. Gerald Massey in his "Natural Genesis," speaking of the origin of the phallic cult, says: "According to Theal, the African historian and collector of folk tales, the Kaffirs have no Sabbath, and keep none of the sacred seasons of periodic recurrence, commonly celebrated by a festival. But from time immemorial, they have preserved the primitive custom of rejoicing at the first appearance of the menstrual period of the female. This they celebrate in what is their sole festival. At that time of a girl's life, all the young women in her neighbourhood meet for rejoicing, at which they celebrate the festival of pubescence. These young women are then distributed among the men who are selected to lie with them, but who are prohibited from sexual intercourse; and if the trespass be committed the men are fined—a primitive mode of paying a price which was afterwards continued in the compensation enforced at the time of marriage. We still keep the birthday and celebrate the coming of age at a fixed period of life; but the festival of puberty is extant to show that the earliest birthday ever memorialized was not the day on which the child was born into the world, but the time of rebirth into womanhood and manhood. When applied to the male, this period of pubescence suggested the birthday of the boy who was at this time admitted as a young man into the *totemic* tribe;— hence the typical 'second birth' celebrated in the mysteries when the first had also been acknowledged.

"It is here we have to seek not only the genesis of time itself, but the origin of the so-called phallic cult, or worship of the generative powers, which did not commence as a religion, but with the sexual typology as a mode of expression, and of keeping time as well as other forms of law. At the age of puberty the boy was first counted as one, an individual, or rather one of the totem; he *counted* because he was reckoned. Until then the children were not reckoned, and did not count either as individuals or members of the totem, consequently they were of no account. *Rekh* (Eg.) is to count and reckon. The *rekh* (Eg.), or *ilk*, as a people of a district or totem were those who were reckoned. To be reckoned and numbered was to be of rank, and this constituted the first honour for the male and the female, even from the time they were reckoned separately in the earliest two castes. Up to this period they were mere slaves, and at puberty they became men and women on whom the freedom of the totem was conferred. This rank and honour of being reckoned as one of the body corporate is shown by the Hottentot language, in which the word Goa, to count, also signifies honour and respect; *Goei* is one; Goab, the number, also means regard, respect, and honour, which originated in becoming one of the number at puberty, those who were of account. If the Hottentot is slighted, he will say indignantly, 'I am not counted,' *i.e.* he is treated as nobody."

"Three phases," says Westropp, "in the representation of the phallus should be distinguished; first, when it was the object of reverence and religious worship; secondly, when it was used as a protecting power against evil influences of various kinds, and as a charm or amulet against envy and the evil eye, as at the postern gate at Alatri and at Pompeii, and as frequently occurs in amulets of porcelain found in Egypt, and of bronze in Italy; thirdly, when it was the result of mere licentiousness and dissolute morals. Another cause also contributed to its reverence and frequent representation—the natural desire of women among all races, barbarous as well as civilized, to be the fruitful mother of children—especially as, amongst some people, women were esteemed according to the number of children they bore, and as, among the Mahommedans of the present day, it is sinful not to contribute to the population; as a symbol, therefore, of prolificacy, and as the bestower of offspring, the phallus became an object of reverence and especial worship among women."

CHAPTER II.

Deities connected with Phallic Worship.

"THE Egyptians in their hymns to Osiris, invoked that god as the being who dwelt concealed in the embraces of the sun; and several of the ancient Greek writers speak of the great luminary itself as the generator and nourisher of all things, the ruler of the world, the first of the deities, and the supreme Lord of all mutable or perishable beings. Not that they any more than the Egyptians, deified the sun considered merely as a mass of luminous or fervid matter; but as the centre or body, from which the pervading spirit, the original producer of order, fertility and organization, amidst the inert confusion of space and matter, still continue to emanate through the system, to preserve the mighty structure which it had formed. The primitive spirit is said to have made the sun to guard and govern all things, it being thought the instrumental cause through which the powers of reproduction, implanted in matter, continued to exist; for without a continued emanation from the active or male principle of generation, the passive or female principle which was derived from it, would of itself become exhausted.

"This continued emanation, the Greeks personified into two distinct personages, the one representing celestial love, or attraction, and the other, animal love or desire, to which the Egyptians added a third, by personifying separately the great fountain of attraction, from which both were derived. All the three were, however, but one, the distinctions arising merely out of the metaphysical subtlety of the theologists, and the extravagant allegories of the poets, which have a nearer resemblance to each other than is generally imagined.

"This productive ethereal spirit being expanded through the whole universe, every part was in some degree impregnated with it, and therefore every part was, in some measure, the seat of the deity, whence local gods and goddesses were everywhere worshipped, and consequently multiplied without end. 'Thousands of the immortal progeny of Jupiter,' says Hesiod, 'inhabit the fertile earth, as guardians to mortal men.' An adequate knowledge, either of the number or attribute of these, the Greeks never presumed to think attainable, but modestly contented themselves with revering and invoking them whenever they felt or wanted their assistance." *

Ceres and Bacchus (or Demeter and Dionysus or Iacthus), called in Egypt, Isis and Osiris, and in Syria, Venus and Adonis (Astartê and Adoni), were the deities in whose names and under whose protection persons were most commonly instructed in the faith. Thus, Herodotus (II. ch. 42), "Such Egyptians as possess a temple of the Theban Jove, or live in the Thebaic canton, offer no sheep in sacrifice, but only goats; for the Egyptians do not all worship the same gods, excepting Isis and Osiris, the latter of whom they say is the Grecian Bacchus."

And Euripides (*Bacchæ* 73), "Oh happy, blessed is he that witnesseth the initiation of the deities, for he venerateth the source of life; not only does he divine the Orgies of Cybele, the Great Mother, but waving the thyrsus, and crowned with ivy, he is also a votary of Dionysus."

The word Bacchus or Iacchus is a title derived from the exclamations uttered in the festivals of this god, whose other Latin name, Liber, is also a title signifying the same attribute as the Greek epithet, Lusios, or Luson. But whence the more common Greek name, Dionusos, is derived, or what it signifies, is not so easy to determine, or even to conjecture with any reasonable probability. The first part of it appears to be from Deus, Dios, or Dis, the ancient name of the supreme universal god; but whether the remainder is significant of the place from which this deity came into Greece, or of some attribute belonging to him, we cannot pretend to say, and the conjectures of etymologists, both ancient and modern, concerning it, are not worthy of notice. An ingenious writer in the *Asiatic Researches* derives the whole name from a sanscrit title of an oriental demi-god, and as Ausonius says it was Indian, this derivation appears more probable than most others of the kind.

At Sicyon in the Peloponnesus, he was worshipped under another title, which we shall not venture to explain any further than that it implies his having the peculiar superintendence and direction of the characteristics of the female sex. (Clement, of Alexandria, declares that he was denominated Choiropsale by the Syconians, a low term expressing immodest practices with

* Knight. *Symb. Lan. Anc. Art.*

women). At Lampascus, too, on the Hellespont, he was venerated under a symbolical form adapted to a similar office, though with a title of a different signification, *Priapus*.

According to Herodotus, the name Dionysus, or Bacchus, with the various obscene and extravagant rites that distinguished his worship, was communicated to the Greeks by Melampus, who appears to have flourished about four generations before the Trojan war, and who is said to have received his knowledge of the subject from Cadmus and the Phœnicians, who settled in Bœotia. "He it was," says Herodotus, "who introduced into Greece the name of Bacchus, the ceremonial of his worship, and the procession of the phallus. He did not, however, so completely apprehend the whole doctrine as to be able to communicate it entirely; but various sages since his time have carried out his teaching to greater perfection. Still it is certain that Melampus introduced the phallus, and that the Greeks learnt from him the ceremonies which they now practise. I therfore maintain that Melampus, who was a wise man, and had acquired the art of divination, having become acquainted with the worship of Bacchus through knowledge derived from Egypt, introduced it into Greece, with a few slight changes, at the same time that he brought in various practices. For I can by no means allow that it is by mere coincidence that the Bacchic ceremonies in Greece are so nearly the same as the Egyptian—they would then have been more Greek in their character, and less recent in their origin. Much less can I admit that the Egyptians borrowed these customs, or any other, from the Greeks. My belief is that Melampus got his knowledge of them from Cadmus the Tyrian, and the followers whom he brought from Phœnicia into the country which is now called Bœotia." *

General tradition has attributed the introduction of the mystic religion into Greece, to Orpheus, a Thracian; who, if he ever lived at all, lived probably about the same time with Melampus, or a little earlier. The traditions concerning him are, however, extremely vague and uncertain, and the most learned and sagacious of the Greeks is said to have denied that such a person had ever existed; but nevertheless, we learn from the very high authority of Strabo that the Greek music was all Thracian or Asiatic, and, from the unquestionable testimony of the Iliad, that the very ancient poet Thamyris was of that country, to which tradition has also attributed the other old sacerdotal bards, Musæus and Eumolpus.

* Rawlinson's *Herodotus*, II., ch. 49.

As there is no mention, however, of any of the mystic deities, nor of any of the rites with which they were worshipped, in any of the genuine parts, either of the Iliad or Odyssey, nor any trace of the symbolical style in any of the works of art described in them, nor of allegory or enigma in the fables which adorn them, we may fairly presume that both the rites of initiation and the worship of Bacchus are of a later period, and were not generally known to the Greeks till after the composition of those poems. The Orphic Hymns, too, which appear to have been invocations or litanies used in the mysteries, are proved, both by the language and the matter, to be of a date long subsequent to the Homeric times, there being in all of them abbreviations and modes of speech not then known, and the form of worshipping or glorifying the deity by repeating adulatory titles, not being then in use, though afterwards common.

In Egypt, nevertheless, and all over Asia, the mystic and symbolical worship appears to have been of immemorial antiquity. The women of the former country carried images of Osiris in their sacred processions, with a movable phallus of disproportionate magnitude, the reason for which Herodotus does not think proper to relate, because it belonged to the mystic religion. Diodorus Siculus, however, who lived in a more communicative age, informs us that it signified the generative attribute. In Book I., chap. 6, this author says: "They say the goat was accounted amongst the number of the Gods, for the sake of his genitals, as Priapus is honoured among the Grecians, for this creature is exceeding lustful, and therefore they say that member (the instrument of generation) is to be highly honoured as that from which all living creatures derive their original. They say that these privy parts are not only accounted sacred among the Egyptians, but among many others are religiously adored in the time of their solemn rites of religious worship, as those parts that are the causes of generation. And the priests who succeed in the office, descended to them from their fathers in Egypt, are first initiated into the service of this god. For this reason the Panes and Satyrs are greatly adored among them, and therefore they have images of them set up in their temples, with their privy parts erected like to the goat, which they say is the most lustful creature in the world. By this representation they would signify their gratitude to the gods, for the populousness of their country."

Plutarch also says that the Egyptian statues of Osiris had the phallus to signify his procreative and prolific power, the extension of which through the three elements of air, earth and water, they expressed by another kind of statue, which was occasionally carried in procession, having a triple symbol of the same attribute. His words are: "They exhibit the statue in human semblance, holding the sexual part prominent as fecundating and nourishing. They display the emblem and carry it around, having the sexual parts threefold."

The Greeks usually represented the phallus alone, as a distinct symbol the meaning of which seems to have been among the last discoveries revealed to the initiated. It was the same in emblematical writings, as the Orphic epithet, Pan-genetor, universal generator, in which sense it is still employed by the Hindus. It has also been observed among the idols of the native Americans and ancient Scandinavians; nor do we think the conjecture of an ingenious writer improbable who supposes that the Maypole was a symbol of the same meaning, and the first of May a great phallic festival both among the ancient Britons and Hindus, it being still celebrated with nearly the same rites. The Greeks changed, as usual, the personified attribute into a distinct deity called Priapus, whose universality was, however, acknowledged to the latest period of heathenism. *

Isis, the peculiar goddess of maternity, is often figured in Roman sculpture, holding up in her hand a comical object, pouch-shaped, exhibiting a triangular orifice. This object some have taken for the Persia plum; much more probably does it represent the female organ, the most natural and expressive symbol of that divinity's peculiar function. In her mystic coffer were carried the distinctive marks of both sexes, the *lingam* and *yoni* of the Hindoos. Their Isis, Parvati, who in this character takes the name of Deva "the goddess" pre-eminently, bears in her hand for distinctive badge the yoni, or bhaga, often a precious stone carved into that shape. Similarly her consort, Siva, carries the lingam or phallus. For example, the Nizam's diamond, the largest stone of its kind known certainly to exist, exhibits evident traces of the native lapidary's clumsy endeavours to reduce the native crystal to the proper shape for the hand of the great goddess. Ugly omen to happen under a female reign, this

---

* See Knight's *Ancient Art and Mythology*.

diamond was accidentally broken in two just before the outbreak of the Sepoy revolt. *

"Priapus," says Smith's mythological dictionary, "a son of Dionysus and Aphrodite. Aphrodite, it is said, had yielded to the embraces of Dionysus, but during his expedition to India, she became faithless to him, and lived with Adonis. On Dionysus' return from India, she indeed went to meet him, but soon left him again, and went to Lampsacus on the Hellespont, to give birth to the child of the god. But Hera, dissatisfied with her conduct, touched her, and, by her magic power, caused Aphrodite to give birth to a child of extreme ugliness, and with unusually large genitals. This child was Priapus. According to others, however, Priapus was a son of Dionysus and a Naiad or Chione, and gave his name to the town of Priapus, while others again describe him as a son of Adonis, by Aphrodite, or as the son of a long-eared father, that is, of Pan or a Satyr. The earliest Greek poets, such as Homer, Hesiod, and others, do not mention this divinity, and Strabo expressly states, that it was only in later times that he was honoured with divine worship, and that he was worshipped more especially at Lampsacus on the Hellespont, whence he is sometimes called Hellespontiacus. We have every reason to believe that he was regarded as the promoter of fertility both of the vegetation and of all animals connected with an agricultural life, and in this capacity he was worshipped as the protector of flocks of sheep and goats, of bees, the vine, all garden produce, and even of fishing. Like other divinities presiding over agricultural pursuits, he was believed to be possessed of prophetic powers, and is sometimes mentioned in the plural. As Priapus had many attributes in common with other gods of fertility, the Orphics identified him with their mystic Dionysus, Hermes, Helios, &c. The Attic legends connect him with such sensual and licentious beings as Conisalus, Orthanes, and Tychon. In like manner he was confounded by the Italians with Mutunus, the personification of the fructifying power in nature. The sacrifices offered to him consisted of the first fruits of gardens, vineyards and fields, of milk, honey, cakes, rams, asses, and fishes. He was represented in carved images, mostly in the form of Hermæ with very large genitals, carrying fruit in his garment, and either a sickle or cornucopia in his hand."

Diodorus says: "The Egyptians tell this story concerning Priapus; they say that the Titanes in ancient times treacherously

* King's *Gnostics*.

assassinated Osiris, and divided the members of his body into equal parts. and that every one privately carried away a part out of the palace, only his privy members they threw into the river, because none would meddle with them. But Isis, they say, after a diligent inquiry made concerning the murder of her husband, and having revenged his death upon the Titans, by conjoining his dismembered parts, reduced them to a human shape, and delivered the body to the priests to be buried, and commanded that Osiris should be adored as a god, and appointed the shape of his privy member (which only was wanting and could not be found) to be set up as a sacred relict in the temple, and to be honoured likewise as a deity; and these are the things which the ancient Egyptians feign concerning the original and divine worship of Priapus. This god is not only honoured in the festivals of Bacchus, but in all other sacred solemnities, where with sport and ridicule his image is presented to the view of all."

Pan, like other mystic deities, was wholly unknown to the first race of poets; there being no mention of him in either the Iliad, the Odyssey, or in the genuine poem of Hesiod; and the mythologists of later times having made him a son of Mercury by Penelope, the wife of Ulysses; a fiction, perhaps, best accounted for by the conjecture of Herodotus, that the terrestrial genealogies of the mystic deities, Pan, Bacchus, and Hercules, are mere fables bearing date from the supposed time when they became objects of worship. Both in Greece and Egypt, Pan was commonly represented under the symbolical form of the goat half humanised; from which are derived his subordinate ministers or personified emanations, called Satyrs, Fawns, Tituri, Paniskoi; who, as well as their parent, were wholly unknown to the ancient poets.

Pan is sometimes represented ready to execute his characteristic office, and sometimes exhibiting the result of it; in the former of which, all the muscles of his face and body appear strained and contracted; and in the latter, fallen and dilated; while in both the phallus is of disproportionate magnitude, to signify that it represented the predominant attribute. In one instance he appears pouring water upon it, but more commonly standing near water, and accompanied by aquatic fowls; in which character he is confounded with Priapus, to whom geese were particularly sacred. Swans, too, frequently occur as emblems of the waters upon coins; and sometimes with the head of Apollo on the reverse, when there may be some allusion to the ancient notion of their singing; a notion which seems to have arisen from the noises which they make in the high latitudes of the North, prior to their departure at the approach of winter.

Though the Greek writers call the deity who was represented by the sacred goat at Mendes, *Pan*, he more correctly answers to Priapus, or the generative attribute considered abstractedly; which was usually represented in Egypt as well as in Greece, by the phallus only. The deity was honoured with a place in most of their temples, as the Lingam is in those of the Hindus; and all the hereditary priests were initiated or consecrated to him, before they assumed the sacerdotal office: for he was considered as a sort of accessory attribute to all the other divine personifications, the great end and purpose of whose existence was generation or production. A part of the worship offered both to the goat Mendes, and the bull Apis, consisted in the woman tendering their persons to him, which it seems the former often accepted, though the taste of the latter was too correct. An attempt seems to have been made, in early times, to introduce similar acts of devotion into Italy; for when the oracle of Juno was consulted upon the long-continued barrenness of the Roman matrons, its answer was "Iliadas matres caper hirtus inito" (Let the rough goat approach the Trojan matrons), but those mystic refinements not being understood by that rude people, they could think of no other way of fulfilling the mandate, than sacrificing a goat, and applying the skin cut into thongs, to the bare backs of the ladies, which however had the desired effect: "Virque pater subito, nuptaque mater erat" (Speedily the man a father, the wife a mother was).

At Mendes, female goats were also held sacred, as symbols of the passive generative attribute; and on Grecian monuments of art, we often find caprine satyrs of that sex. The fable of Jupiter having been suckled by a goat, probably arose from some emblematical composition, the true explanation of which was only known to the initiated. Such was Juno Sospita of Lanuvium, near Rome, whose goat-skin dress signified the same as her title; and who, on a votive car of very ancient Etruscan work found near Perugia, appears exactly in the form described by Cicero, as the associate of Hercules dressed in the lion's skin, or the Destroyer. *

* Knight's *Ancient Art.*

CHAPTER III.

Emblems connected with Phallic Worship.

IT was the opinion of the ancients, that all the constituent parts of the great machine of the universe were mutually dependent upon each other; and the luminaries of heaven, while they contributed to fecundate and organise terrestrial matter, were in their turn nourished and sustained by exhalations drawn from the humidity of the earth and its atmosphere. Hence the Egyptians placed the personifications of the sun and moon in boats (Plutarch: Isis and Osiris), while the Greeks, among whom the horse was a symbol of humidity, placed them in chariots, drawn sometimes by two, sometimes by three, and sometimes by four of these animals; which is the reason of the number of Bigæ, Trigæ, and Quadrigæ, which we find upon coins: for they could not have had any reference to the public games, as has been supposed, a great part of them having been struck by states, which not being of Hellenic origin, had never the privilege of entering the lists on those occasions. The vehicle itself appears likewise to have been a symbol of the female generative power, or the means by which the emanations of the sun acted; whence the Delphians called Venus by the singular title of the Chariot (Plutarch); but the same meaning is more frequently expressed by the figure called a Victory accompanying it; and by the fish, or some other symbol of the waters, under it. In some instances, have been observed, composite symbols signifying both attributes in this situation; such as the lion destroying the bull, or the Scylla, which is a continuation of emblems of the same kind, as those which compose the Sphinx and Chimæra, and has no resemblance to the monster described in the Odyssey. *

Ancient writers describing the initiation of votaries into the celebrated mysteries of Greece state that just when the ceremonies were being brought to a close, the symbolic image of the fecundity of nature was then exhibited; an image that expressed the means by which she renews herself in the class of organised

* Knight. *Sym. Lan. Anc. Art.*

bodies, and which, having been at first chosen by a simple and rude people, had continued in use after they were civilized and corrupted, because it had been originally consecrated to religious purposes. The *Phallus* was carried in great pomp; in the ceremonies of the women, the Kteis was made use of; and in spite of the remonstrances of the fathers of the church, it would appear that this ceremony still continued to be respected. But it conveyed no impure idea to the imagination, for the initiated addressed this prayer to nature.

"Hail! holy and unwearied benefactress of the human race! thou who, like a tender mother, lavishest on mortals thy precious gifts, and who stretchest forth thy hands to assist the unhappy, all hail! I invoke thee, thou powerful deity; thee, whom the gods of heaven adore, and whom the gods of hell dread: thee, who hast impressed motion on the celestial spheres; who continuest to nourish the fires of the sun; who governest the universe; and whose empire extends even to Tartarus. Thou speakest and the stars make answer; the gods rejoice, the seasons succeed each other, and the elements are obedient to thy voice. By thy order the winds rage, and clouds are collected; plants germinate and issue from the bosom of the earth; animals people the forests and the mountains; the serpent hides himself in obscure retreats; the inhabitants of air, the monsters of the ocean, the whole universe is subject to thy command. Who can worthily celebrate thy praises, O august divinity! Engrossed with thy majesty, I shall incessantly behold thee and contemplate thy divine perfections. May the sacred image never cease to dwell in the bottom of my heart."

The reverence paid to fish of different kinds by the Egyptians and some other ancient nations was very marked. Historians say that all the natives of the river were in some degree esteemed sacred. In many parts the people did not feed upon them. The priests in particular never tasted fish; and this on account of their imputed sanctity, for they were sometimes looked upon as sacred emblems: at other times worshipped as real deities. One species of fish was styled Oxurunchus; and there was a city of the name, built in honour of it, and a temple where this fish was publicly worshipped. Nor was the veneration confined to this place, but prevailed in many other parts of Egypt. A fish called Phagrus was worshipped at Syene: as the Mæotis was at Elephantis. The Lepidotus had the like reverence paid to it: as had also the Eel; being each sacred to the god Nilus. This is ridiculed in a passage, which has been often quoted, from the ancient comedian Antiphanes: who mentions, that an eel by the Egyptians was reverenced equally with the gods. Another comedian says that they esteemed it as one of their supreme deities: and he at the same time exposes their folly with some humour. A Grecian is made to address himself to an Egyptian: and he accordingly says,—" It is impossible for me to ride in the same troop with you: for our notions and manners are diametrically opposite. You pay adoration to an ox: I kill and sacrifice it to the gods. You esteem an eel to be a very great divinity. I only think it the best dish that comes upon the table. You worship a dog. I whip him handsomely; especially if I find the cur purloining my dinner." *

Here it is proper to take notice, that there was a female deity called Athor in Egypt: but in Syria Atar-Cetus, or Atargatis; and abbreviated, Dercetus and Derceti. This personage was supposed to have been of old preserved by means of a fish: and was represented one half under that form; and the other half as a woman. She was esteemed to be the same as the Aphrodite of the Greeks and the Venus of the Romans, whose origin was from the sea. In consequence of this, wherever her worship prevailed, fish were esteemed sacred; and the inhabitants would not feed upon them. This was the case at Edessa, called Hierapolis, where Atargatis or Derceto was held in particular veneration. Xenophon in his march through these parts observed, in a river called Chalus, many large fishes, which appeared tame, and were never taken for food: the natives esteeming them as gods. Lucian tells us, that this worship was of great antiquity; and was introduced into these parts from Egypt. The same custom seems to have been kept up in Babylonia: but what was of more consequence to the Israelites, it prevailed within their own borders. Dagon of Asdod, or Azotus, was the same deity: and represented under a like figure as Atargatis. The same rites and abstinence were observed also at Ascalon. Diodorus Siculus speaks of this city, which he places in Syria, rather than Palestine; at no great distance from which he says was a large lake, abounding with fishes. Near it was a noble temple of the goddess Derceto, whom they represented with the face of a woman, but from thence downwards, under the figure of a fish. The history of Derceto in this place was, that she threw herself into this

* *Anaxandrides.*

lake, and was changed to a fish. On which account the inhabitants of Ascalon, and of some parts of Syria, abstained from fish: and held those of the lake as so many deities. However strange this idolatry may appear, yet we see how very far it reached; and with what a reverence it was attended. It was to be found not only in Syria, which was sufficiently near, but in the borders of Lebanon; also at Ascalon, Ashdod, and Joppa; which cities were within the precincts of the tribes of Dan and Judah. *

The Egyptians honoured the Nile with a religious reverence, and valued themselves much upon the excellence of their river. Nor was this blind regard confined to the Egyptians only, but obtained in many parts of the world. Herodotus says of the Persians, that of all things rivers were held in the highest veneration. They worshipped them, and offered to them sacrifices: nor would they suffer anything to be thrown into them that could possibly pollute their waters. The like obtained among the Medes, Parthians, and the Sarmatians. We read in Homer of the sanctity in which rivers were held in Greece. Among these more especially were the Spercheius, Peneus, Achelous, and Alpheus. The last had altars, and sacrifices offered to him in common with Diana. The Phrygians made the like offerings to the Marsyas and Mæander.

But no nation carried their reverence to such an extravagant degree of idolatry as the Egyptians. They looked upon their river not only as consecrated to a deity: but if we may believe some authors, as their chief national god, and worshipped it accordingly. The people above Syene, styled the Nile Siris and Sirius, which was the name of Osiris, and the Sun: and upon solemn occasions made invocations to it as their chief guardian and protector. They supposed that it gave birth to all their deities, who were born upon its banks, and that it was particularly the father of Vulcan. Hence there were temples erected to his honour, and a city called after his name, Nilopolis; in which he was particularly worshipped: and there were festivals and rites, styled Neiloa Sacra, which were observed all over Egypt. As they received so much benefit from their river, they held water in general sacred, as Julius Firmicus has observed.

These superstitions, and this veneration for the river prevailed, as we may presume, even in the time of Moses. This may be inferred from the like notions being found in the earliest

---

* Bryant on the Plagues of Egypt.

ages among the Syrians and Babylonians. The same prevailed in Greece. They were brought over to the last region by colonies from Egypt, and appear to have been of very early date. The ancient Grecians supposed many of their kings and heroes to have been the offspring of rivers: and the Sea, or Oceanus, was esteemed the father of their gods. This was borrowed from Egypt, for the natives of that country esteemed the Nile to be the ocean. *

"Loss of hair was degrading and humiliating, whether voluntary or enforced, and shaving is the symbolic act of rendering non-virile, monkish, unsexual, whether applied to the pubes, beard, or crown, as it was in Egypt, and still is in the Cult of the Virgin Mother and her impubescent Bambino in Rome.

This is recognised by Isaiah, who threatens Israel with a razor that will shave it at both ends, and '*it shall consume the beard*' (ch. vii. 20).

As hair was the emblem of virility and reproduction, baldness was the natural antithesis; and the loss of the hair was enforced as a later form of penalty, because it had been held so sacred as a voluntary offering. The hair being a symbol of reproducing potency, this will account for the lock of a person's hair being considered the representative of the person's self, when his life is sought to be taken, or blasted by magic, *i.e.* enacting of the malignant desire in gesture-language according to primitive usage.

It is believed that the hair and nails ought never to be cut on Sunday, the day of Khem-Horus,† or on Friday, the day of the Genitrix.

The Lion Paru in the Ritual is called the 'Lord of numerous transformations of skins,' *i. e.*, repeatings of the hair; and time was, in England, when people would make a point of having their hair cut whilst the moon, the female reproducer, was in the sign of the Lion or the Ram, two chief types of male potency.

When we know the symbolic value of nail from the origin, we can understand the reason why biting the nail by way of scorn should be considered an insult. The act was equal to plucking the beard or cutting the hair; it was aimed at the person's

---

* Bryant on the Egyptian Plagues.

† There is a like superstition lingering to the present day in England embodied in the following couplet:—
"He that cutteth hair or horn
Shall rue the day that he was born."

manhood, on the ground of nail being a representative of virility in gesture-language and the primitive typology.

The nails as an equivalent for the hair, a type of '*renewal coming of itself*," will account for a custom like this:—'The ancient Frenchmen had a ceremony that when they would marry, the bridegroom should pare his nails and send them to his new wife; which done, they lived together afterwards as man and wife.' The act had the same significance as when the pubes or locks of hair were offered to the divine Genitrix, or the foreskins were piled in the circle of the twelve stones at Gilgal. Each was dedicated to reproduction.

Captain Cook describes the Maori as wearing the nails and teeth of their dead relatives. These were equivalent to the phallus worn by the widows, as a type of reproduction.

It was an Egyptian custom to gild the nails, teeth, and membrum virile of the embalmed mummy. These were glorified in the gloom of the grave because, as types of production, they served in a second place as emblems of foundation, and visible basis of renewal and resurrection.

It was a theory that the hair, beard, and nails of the Japanese Mikado were never cut. They had to be trimmed furtively while he was sleeping. This corresponds to the assumption that the king never dies. He was not reproducible. He was only transformed. He was the living one, like the Aukh; an image of the ever-being, a type of the immortal.

The male emblem of virility, like the scalp, was a trophy to be cut off in battle. On the monuments there are heaps of these collected as evidence of conquest. In one instance the 'spoils of the Rebu,' consist of donkey-loads of phalluses (Karanatu) and several hands. Twelve thousand, five hundred and thirty-five members and hands were cut off from the dead after the battle of Khesef-Tamahu, and deposited as proofs of victory—an enacted report—before the Pharaoh Rameses.

By the aid of the hieroglyphic values conferred on the image in life, we can read the significance of the emblem in death. By its excision the enemy were typically annihilated; the last tribute paid thus was the forfeiture of his personality in a spiritual sense; for without the member, the deceased, according to Egyptian thought, could not be reconstructed. He would not rise again; resurrection, as in the case of Osiris, depended on repossessing the member. The type of individuality here was the emblem of existence hereafter.

We have only to become acquainted with the doctrines of the mummy in the Ritual, and see the fearful anxiety of the deceased to get all his members intact and solid, to avoid dissolution ; see how he rejoices in the firmness of his phallus, the hardness of his heart, the soundness and indissolubility of his vertebræ, to apprehend what terrible meaning there was in the custom of dismembering the body, swallowing the eyes, eating the heart, or pulverising the bones to drink them in water as an ocular demonstration of dissolution. The New Zealanders are said to think that a man who is eaten is thus destroyed soul and body.

In the Atharva-Vada it is affirmed that when the dead passed through the sacrificial fire of heaven, Agni (fire) does not consume their generative organ ; whereas in the earlier thought of Kam it would have been held to do so, or to efface the type, which came to the same thing, symbolically, as the most physical plane of thought.

Because the custom was typical, it permitted of modification and commutation in the interchange of types. Thus the '*bloody foreskin*' of the slain came to be adopted in place of the total emblem, as with the Abyssinians, described by Bruce, and the hundred Philistine foreskins demanded by Saul of David, and doubled as the dowry of Michal. The foreskin, or prepuce cover had precisely the same symbolical value as the sign of manhood, hence the excision at the age of puberty, for that was the earlier period, and the Jewish custom does not retain the primary significance, except in its being a commutated offering to the paternal deity."*

"It was stated in the *Paris Moniteur*, during the month of January, 1865, that in the province of Venice, Italy, excavations of a bone-cave were made, and bones of animals, chiefly posttertiary, were found together with flint implements, a needle of bone, having both eye and point, and a plate of an argillaceous compound, on which was scratched a rude figure of the male organ of generation ; and that these things were dug from beneath ten feet of stalagmite. That emblem was a type of resurrection, formed on the most natural grounds. According to the Gnosis, this rude figure had the same significance, denoting a place of burial for those who expected to rise again, and its image in the tomb can be read by the Egyptian 'Litany of Ra' (34). 'Homage to thee, Ra ! supreme power, the way of light in the sarcophagus ! Its form is that of the progenitor.'

* Massey's *Natural Genesis*.

The self-erecting member was the type of resurrection, as the image of Khem-Horus, the re-arising sun, and of Khepr-Ra, the re-erector of the dead. The widows of the aborigines of Australia are in the habit of wearing the dead husband's phallus round their necks, and the significance of the custom is the same as in Egypt and the bone-caves. The emblem was sacred as the type of reproduction. The same type was worn as an ear-drop by the ladies of Latium, and is yet worn in Southern Italy.

'*Images of pollution have been found at Hissarlik*,' exclaims the author of *Juventus Mundi*, and the voice of the primitive consciousness says the phallus typified the earliest ray of light that penetrated the darkness of the grave ; indeed this primitive type is found in a fourfold form in the Christian iconography of the Roman catacombs.

The branch of palm has now taken its place in the imagery of heaven and the typology of the eternal. In the Book of Revelation those who stand before the throne are pourtrayed with palms in their hands. Horus is represented in the monuments as defending himself against his evil enemy, Sut, or Satan, with a palm-branch in his hand, The branch of palm was, and still is an emblem of renewal. But the branch of birch that was buried with the dead in the barrows had the same meaning. A barrow at Kepwick was found to be lined with the bark and branches of the birch. That is the Bedmen of the British, which was also the maypole and the phallus. The Bedmen was typical of the resurrection equally with the palm.

The beetle type of Kephr, the transformer, was also buried with the British dead as with the Egyptian, likewise also beads, as with the Africans and Egyptian mummies. As these were imperishable it should be noticed that a kind of bead which is made in Africa has been found buried in Britain. Beads denote reproduction, and were worn by the genitrix Isis when enceinte, as the beads and berries are worn by the pregnant women in Africa to-day. Beads in the tombs typified re-birth, whether in Africa, America, Australia, or Britain."*

The most celebrated of all the sacred animals worshipped by the Egyptians was Mnevis or Apis, the bull, the Epaphus of the Greeks, whose image is found to an enormous extent marked upon the coins and religious monuments of various nations. Strength was what it was desired to express, and the bull being

* Massey's *Natural Genesis*.

the most powerful animal known in climates too cold for the propagation of the elephant, was the creature adopted. It was under this form that the mystic Bacchus, or generative power was represented, not only upon the coins, but in the temples of the Greeks. It was sometimes nothing but a bull; at others, a bull with a human face; and at others entirely human with the exception of the horns and ears; in some instances also the head was adorned with a beard.

According to Plutarch and Herodotus, the Mnevis of the Egyptians was held by some to be the mystic father of Apis; and as the one has the disk upon the head, and was kept in the city of the sun, while the other is distinguished by the crescent, it is probable that the one was the emblem of the divine power acting through the sun; and the other, of it acting through the moon, or (what was the same) through the sun by night. Apis, however, held the highest rank, he being exalted by the superstition of that superstitious people into something more than a mere symbol, and supposed to be a sort of incarnation of the deity in a particular animal, revealed to them at his birth by certain external marks, which announced his having been miraculously conceived by means of a ray from heaven. Hence, when found, he was received by the whole nation with every possible testimony, of joy and gratulation, and treated in a manner worthy of the exalted character bestowed on him; which was that of the terrestrial image or representative of Osiris, in whose statues the remains of the animal symbol may be traced.

Their neighbours the Arabs appear to have worshipped their god under the same image, though their religion was more simple and pure than that of any heathen nation of antiquity, except the Persians, and perhaps the Scythians. They acknowledge only the male and female, or active and passive powers of creation, the former of whom they called Urotalt, a name which evidently alludes to Urus. Herodotus calls him Bacchus, as he does the female deity, Celestial Venus, by which he means no more than than they were personifications of the attributes which the Greeks worshipped under those titles.

In most Greek and Roman statues of the bull, whether in the character of Mnevis or Apis, of both of which many are extant of a small size in bronze, there is a hole upon the top of the head between the horns where the disk or crescent, probably of some other material, was fixed: for as the mystical or symbolical was engrafted upon the old elementary worship, there is always a link of connection remaining between them. The Bacchus of the Greeks, as well as the Osiris of the Egyptians, comprehended the whole creative or generative power, and is therefore represented in a great variety of forms, and under a great variety of symbols, signifying his subordinate attributes. *

Priapus was celebrated by the Greek poets, under the title of Eros, Love or Attraction, the first principle of animation, the father of gods and men, and the regulator and disposer of all things. He is said to pervade the universe with the motion of his wings, bringing pure light: and thence to be called the splendid, the self-illumined, the ruling Priapus—light being considered in this primitive philosophy as the great nutritive principle of all things. Wings are attributed to him as the emblems of spontaneous motion; and he is said to have sprung from the egg of night, because the egg was the ancient symbol of organic matter in its inert state, or, as Plutarch calls it, the material of generation, containing the seeds and germs of life and motion without being actually possessed of either. It was therefore carried in procession at the celebration of the Mysteries; for which reason Plutarch declines entering into a more particular disquisition concerning its nature, the Platonic interlocutor in the Dialogue observing, that, though a small question, it comprehended a very great one, concerning the generation of the world itself, known to those who understood the Orphic and sacred language, the egg being consecrated, in the Bacchic mysteries, as the image of that which generated and contained all things in itself. His words are:—" They suspected that I held the Orphic and Pythagorean dogmas, and refused to eat the egg (as some do the heart and brain) because it is sacred; imagining it to be the first principle of generated existence.* * * Soon after Alexander proposed the problem concerning the egg and the bird, which was the first. My friend Sylla saying that with this little question, as with an engine, was invoked the great and weighty one concerning the genesis of the world, declaring his dislike of such problems. * * * I speak to those who understand the sacred legend of Orpheus, which shows not only that the egg is before the bird, but makes it before all things. The other matter we will not speak about, being as Herodotus says, of a mystic character. * * * Therefore in the Orgies of Dionysus it is usual to consecrate an egg as representing that which generates and contains all things in itself."

* See Knight's *Symbolical Language of Ancient Art*.

Venus-Urania, the Mother-Goddess. The characteristic attribute of the passive generative power was expressed in symbolical writing, by different enigmatical representatives of the most distinctive characteristic of the female sex; such as the shell, or concha veneris, the fig-leaf, barley corn, or the letter delta; all which occur very frequently upon coins and other ancient monuments in this sense. The same attribute, personified as the goddess of love or desire, is usually represented under the voluptuous form of a beautiful woman, frequently distinguished by one of these symbols, and called Venus, Kypris, or Aphroditê, names of rather uncertain etymology. She is said to be the daughter of Jupiter and Dionê; that is, of the male and female personifications of the all-pervading spirit of the universe; Dionê being, as before explained, the female Dis or Zeus, and therefore associated with him in the most ancient oracular temple of Greece at Dodona.

Mr. R. P. Knight in his "Symbolical Language of Ancient Art" quotes the following authorities relating to the foregoing.

Clement of Alexandria: *Exhortations*. "The Kteis gunakeios (woman's comb), which is, to speak with a euphemism, and in mystic language, the female sexual parts."

Plutarch: *Isis and Osiris*. "They make a figure of a fig-leaf both for the king and soutnern climate, which fig-leaf is interpreted to mean the generating and fecundating of the universe, for it seems to have some resemblance to the sexual parts of a man."

Eustathius: *on Homer*. "The barley-corn, denoting the vulva among the writers upon the Bacchic Komuses."

Suidas: "Delta, the fourth letter; it also signifies the vulva."

The Genetullides or Genaidai, were the original and appropriate ministers and companions of Venus, who was, however, afterwards attended by the Graces, the proper and original attendants of Juno; but as both these goddesses, were occasionally united and represented in one image, the personifications of their respective subordinate attributes might naturally be changed. Other attributes were on other occasions added, whence the symbolical statue of Venus at Paphos had a beard, and other appearances of virility, which seems to have been the most ancient mode of representing the celestial as distinguished from the popular goddess of that name; the one being a personification of a general procreative power, and the other only of animal desire or concupiscence. The refinement of Grecian art, however, when advanced to maturity, contrived more elegant modes of distinguishing them; and in a celebrated work of Phidias, we find the former represented with her foot upon a tortoise, and in a no less celebrated one of Scopas, the latter sitting upon a goat. The tortoise, being an androgynous animal, was aptly chosen as a symbol of the double power, and the goat was equally appropriate to what was meant to be expressed in the other.

The same attribute was on other occasions signified by the dove or pigeon, by the sparrow, and perhaps by the polypus, which often appears upon coins with the head of the goddess, and which was accounted an aphrodisiac, though it is likewise of the androgynous class. The fig was a still more common symbol, the statues of Priapus being made of the tree, and the fruit being carried with the phallus in the ancient processions in honour of Bacchus, and still continuing among the common people of Italy to be an emblem of what it anciently meant: whence we often see portraits of persons of that country painted with it in one hand, to signify their orthodox devotion to the fair sex. Hence, also, arose the Italian expression, *far la fica*, which was done by putting the thumb between the middle and fore fingers, as it appears in many Priapic ornaments now extant; or putting the finger or the thumb into the corner of the mouth, and drawing it down, of which there is a representation in a small Priapic figure of exquisite sculpture engraved, among the antiquities of Herculaneum.

The myrtle was a symbol both of Venus and Neptune, the male and female personifications of the productive powers of the waters, which appears to have been occasionally employed in the same sense as the fig and fig-leaf, but upon what account it is not easy to guess. Grains of barley may have been adopted from the stimulating and intoxicating quality of the liquor extracted from them, or, more probably, from a fancied resemblance to the object, which is much heightened in the representations of them upon some coins where they are employed as accessory symbols in the same manner as fig-leaves are upon others. Barley was also thrown upon the altar, with salt, the symbol of the preserving power, at the beginning of every sacrifice, and thence denominated *oulochutai*. The thighs of the victim, too, were sacrificed in preference to every other part, on account of the generative attribute, of which they were supposed to be the seat, whence, probably, arose the fable of Bacchus being nourished and matured in the thigh of Jupiter.*

* Symbolical language of Ancient Art.

Humidity in general, and particularly the Nile, was called by the Egyptians the outflowing of Osiris ; (Plutarch) who was with them the God of the Waters, in the same sense as Bacchus was among the Greeks ; whence all rivers, when personified, were represented under the form of the bull ; or at least with some of the characteristic features of that animal. Plutarch says :—"The more learned in arcane matters among the priests, not only term the Nile Osiris, and the sea Typhon, but they also regard Osiris to signify every principle and potency of moisture, venerating it as the cause of generation and the substance of the semen. But by Typhon they mean everything dried, fire-like, and withered, as being opposed to moisture." And again :—"The Greeks consider Dionysus not alone as the patron of wine, but also of the entire moist or generative principle in nature."

In the religion of the Hindus this article of ancient faith, like most others, is still retained ; as it appears from the title, Daughter of the Sun, given to the sacred river Yamuna or Jumna. The God of Destruction is also mounted on a white bull, the sacred symbol of the opposite attribute, to show the union and co-operation of both. The same meaning is more distinctly represented in an ancient Greek fragment of bronze, by a lion trampling upon the head of a bull, while a double phallus appears behind them, and shows the result. The title $\Sigma \omega \tau \eta \rho \ K o \sigma \mu o \upsilon$, upon the composite Priapic figure, published by La Chausse, is well known ; and it is probable that the ithy-phallic ceremonies, which the gross flattery of the degenerate Greeks sometimes employed to honour the Macedonian princes, had the same meaning as this title of Saviour, which was frequently conferred upon, or assumed by them. It was also occasionally applied to most of the deities who had double attributes, or were personifications of both powers ; as to Hercules, Bacchus, Diana, &c.

The head of Proserpina, appears in numberless instances, surrounded by dolphins ; and upon the very ancient medals of Sidè in Pamphylia, the pomegranate, the fruit peculiarly consecrated to her, is borne upon the back of one. By prevailing upon her to eat of it, Pluto is said to have procured her stay during half the year in the infernal regions ; and a part of the Greek ceremony of marriage consists, in many places, in the bride treading upon a pomegranate. The flower of it is occasionally employed as an ornament upon the diadem of both Hercules and Bacchus, and likewise forms the device of the Rhodian medals ; on some of which may be seen distinctly represented an ear of barley springing from one side of it, and the bulb of the lotus, or *Nymphœa nelumbo*, from the other. It therefore holds the place of the male, or active generative attribute; and accordingly we find it on a bronze fragment, published by Caylus, as the result of the union of the bull and the lion, exactly as the more distinct symbol of the phallus is in a similar fragment above cited. The pomegranate, therefore, in the hand of Proserpina or Juno, signifies the same as the circle and cross in the hand of Isis; which is the reason why Pausanias declines giving any explanation of it, lest it should lead him to divulge any of the mystic secrets of his religion.

Inman in his Vocabulary (Ancient Faiths) Article Rimmon, says:—"A pomegranate." The shape of this fruit resembles that of the gravid uterus in the female, and the abundance of seed which it contains makes it a fitting emblem of the prolific womb of the celestial mother. Its use was adopted largely in various forms of worship. It was united with bells, in the adornment of the robes of the Jewish high priest. It was introduced as an ornament into Solomon's temple, where it was united with lilies, and probably with the lotus. In one part of Syria, it was deified, and a temple errected in its honour."

"The arcane meaning of the pomegranate is evidently sexual. The goddess Nana ate of one, and became pregnant. Women celebrating the Thesmorphic, abstained from the fruit rigidly. The Greek name of this fruit, *rhoia*, is a pun for Rhea, the Mother-Goddess. In the phallic symbolism, generation is a part of the mystery of death, and therefore its symbol, the pomegranate, belongs very appropriately to the Queen of the Underworld, who is after all but Isis, Rhea, and Cybelê."

Likewise Pausanias: (Corinth. xvii., 4). "The agalma of Hera is sitting upon a throne, and is of gold and ivory, the work of Polycleitus; her crown has inwrought upon it the Graces and the Hours; in one hand she holds a pomegranate, and in the other, a sceptre; concerning the pomegranate, I will not speak, for it is a matter pertaining to the arcane learning of the mysteries."

Hippocrates says: "All living creatures, not only the animals, but likewise man, originate from the two principles, differing in potency, but agreeing in purpose: I mean fire and water. Fire is able to give life to all things, but water can nourish them." "The soul moveth itself in man, being the commixture of fire and water, necessary to the human body."

The elements—so called, years ago—were supposed to be those in which the active and passive productive powers of the universe respectively existed; since nothing appeared to be produced without them; and whenever they were joined there was production of some sort, either vegetable or animal. Hence they were employed as the primary symbols of these powers on numberless occasions. Among the Romans, a part of the ceremony of marriage consisted in the bride's touching them as a form of consecration to the duties of that state of life upon which she was entering.

"Why do they direct the bride to touch fire and water?" asks Plutarch, "Is it not because, as among the elements and principles, the one is male and the other is female: the one constitutes the principle of motion, and the other the potency existing in matter!"

The Roman sentence of banishment, was an interdiction from fire and water, which implied an exclusion from any participation in those elements, to which all organised and animated beings owed their existence.

According to Plutarch, Numa consecrated the perpetual fire, as the first of all things, and the soul of matter, which, without it, is motionless and dead. Fires of the same kind were, for the same reasons, preserved in most of the principal temples both Greek and Barbarian; there being scarcely a country in the world, where some traces of the adoration paid to it are not to be found. The Prytania of the Greek cities, in which the Supreme Councils were usually held, and the public treasures kept, were so called from the sacred fires always preserved in them. Even common fires were reputed holy by them; and therefore carefully preserved from all contagion of impiety. After the battle of Platæa, they extinguished all that remained in the countries which had been occupied by the Persians, and rekindled them, according to the direction of the Oracle, with consecrated fire from the altar at Delphi. A similar prejudice still prevails, or did till lately, among the native Irish, who anually extinguish their fires, and rekindle them from a sacred bonfire. Perpetual lamps are kept burning in the inmost recesses of all the great pagodas in India; the Hindus holding fire to be the essence of all active power in nature. At Sais in Egypt, there was an annual religious festival called the Burning of Lamps; and lamps were frequently employed as symbols upon coins by the Greeks, who also kept them burning in tombs, and sometimes swore by them, as by known emblems of the deity. The torch held erect, as it was by the statue of Bacchus at Eleusis, and as it is by other figures of him still extant, means life ; while being reversed, as it frequently is upon sepulchral urns and other monuments of the kind, invariably signifies death or extinction.

Though water was thought to be the principle of the passive, as fire was of the active power ; yet both being esteemed unproductive when separate, both were occasionally considered as united in each. Hence Vesta, whose symbol was fire, was held to be equally with Ceres a personification of the Earth, or rather of the genial heat which pervades it, to which its productive powers were supposed to be owing.*

At Mendes a living goat was kept as the image of the generative *power*, to whom the women presented themselves.* * * Herodotus saw this act openly performed, and called it a prodigy. But the Egyptians had no such horror of it ; for it was to them a representation of the incarnation of the Deity, and the communication of his creative spirit to man. It was one of the sacraments of that ancient church, and was, without doubt, beheld with that pious awe and reverence with which devout persons always contemplate the mysteries of their faith, whatever they happen to be, for, as the learned and orthodox Bishop Warburton says, from the nature of any action morality cannot arise, nor from its effects ; therefore, for aught we can tell, this ceremony, however shocking it may appear to modern manners and opinions, might have been intrinsically meritorious at the time of its celebration, and afforded a truly edifying spectacle to the saints of ancient Egypt. Indeed, the Greeks do not seem to have felt much horror or disgust at the imitative representation of it, whatever the historian might have thought proper to express at the real celebration. Several specimens of their sculpture in this way have escaped the fury of the reformers, and remained for the instruction of later times. One of those found among the ruins of Herculaneum, and kept concealed in the Royal Museum of Portici, is well known ; another exists in the collection of Mr. Townley. It may be remarked, that in these monuments the goat is *passive* instead of *active ;* and that the *human symbol* is represented as incarnate with the *divine*, instead of the *divine* with the *human* : but this is in fact no difference ; for the Creator being of both sexes, is represented indifferently

* Knight. *Symb. Lang. Anc. Art.*

of either. In the other symbol of the bull, the sex is equally varied; the Greek medals having sometimes a bull, and sometimes a cow, which Strabo tells us was employed as the symbol of Venus, the passive generative power, at Momemphis, in Egypt. Both the bull and the cow are also worshipped at present by the Hindoos, as symbols of the male and female, or generative and nutritive powers of the Deity. The cow is in almost all their pagodas; but the bull is revered with superior solemnity and devotion.

In the gallery at Florence is a colossal image of the organ of generation, mounted on the back parts of a lion, and hung round with various animals. By this is represented the co-operation of the creative and destroying powers, which are both blended and united in one figure, because both are derived from one cause. The animals hung round show likewise that both act to the same purpose, that of replenishing the earth, and peopling it with still rising generations of sensitive beings. The Chimæra of Homer, of which the commentators have given so many whimsical interpretations, was a symbol of the same kind, which the poet probably having seen in Asia, and not knowing its meaning (which was only revealed to the initiated) supposed to be a monster that had once infested the country. He describes it as composed of the forms of the *goat*, the *lion*, and the *serpent*, and breathing *fire* from its mouth. These are the symbols of the *creator*, the *destroyer*, and the *preserver*, united and animated by *fire*, the divine essence of all *three*. On a gem, published in the Memoirs of the Academy of Cortona, this union of the destroying and preserving attributes is represented by the united forms of the lion and serpent crowned with rays, the emblems of the cause from which both proceed. This composition forms the Chnoubis of the Egyptians.

CHAPTER IV.
Phallic Objects.

THE subjects dealt with in our volume on "Archaic Cup and Ring Marks" have in the minds of many competent antiquarians an important connection with the topics now under discussion. Some of those markings—particularly those occuring in certain parts of India, though it must be admitted that their almost exact counterparts are found elsewhere as well—appear to have a most striking resemblance to some of the monuments and objects which are undoubtedly connected with the ancient worship of the phallus. Speaking of the possible significance of some of the markings, Mr. H. Rivett-Carnac says in his "Archæological Notes on Ancient Sculpturings:"—"It will hardly be contested that some of them bear a striking resemblance to the Mahádeo and yoni marks on the Chandeshwar shrines. The centre mark would appear to do duty for the lings, the circle for the yoni—and the 'gutter' is the depression to be found on most stone yonis, by means of which the votive libations are drained off from the symbols." (The allusion here is chiefly to those sculpturings which consist of a central cup, with a ring outside circular at one part, but running off in a kind of channel at another—pear shaped in fact.) "And here it may be noticed," continues Mr. Carnac, "that in Mahádeo worship the offering of flowers and the pouring of a libation, generally of Ganges water, over the symbols is, so far as I have seen, very general. Those who have visited Benares will remember the little spoons resembling somewhat our Apostles' spoons, some of them beautifully chased with a figure or cobra at the upper end of the handle, used by pilgrims and worshippers at that city in sprinkling the holy water over the Mahádeos there. In Kamám little niches are to be noticed in Mahádeo temples with stone receptacles for holy water, not unlike what are seen in churches abroad. And the temple at Baijnáth boasts of a large well-carved figure holding a bowl, which the priest informed me, held Ganges water, and from which pilgrims sprinkle the Mahádeo placed close by.

"Then again, in connection with the monolith Mahádoes found at Chandeshwar, Pandukoli and Lodh, it may be worth noticing that circles, and what I will call the 'conventional symbols' of the Mahádeo are yoni, and found on exactly similar monoliths in Europe.

"In India these monoliths are found in the centre and in proximity to shrines bearing these markings. Sometimes a circle is found cut on them, and again the shape of the place of worship at Pandukoli, with its double circle of stones in the centre, of the inner of which are the Mahádeos, is as nearly as possible exactly that of these conventional markings.

"I am aware that the view of these markings having reference to lingam worship is not now advanced for the first time. The subject is alluded to at page 93 of Sir J. Simpson's work (on Rock Markings), but only to be summarily dismissed with the following brief remark :—'Two archæological friends of mine, dignitaries in the Episcopal Church, have separately formed the idea that the lapidary cups and circles are emblems of old female lingam worship, a supposition which appears to me to be totally without any anatomical or other foundation, and one altogether opposed by all we know of the specific class of symbols used in that worship, either in ancient or modern times.'"

"I am sanguine, however," continues Mr. Carnac, "that if Sir J. Simpson had seen the sketches of what I have called the 'conventional symbols' on the shrines at Chandeshwar, and had been able to compare them with some of the types figured in his work, he might have been inclined to modify the opinion above extracted. The treatment of these symbols is purely conventional; they bear no anatomical resemblance to anything; they are unlike many of the large, well-known and acknowledged representations of the Mahádeo and yoni. Still, they nevertheless represent the same idea. And here it may be noticed that the same argument of anatomical non-resemblance might be advanced in regard to the well-known representations, common throughout India, of the meaning of which to the initiated there is no doubt at all. To the uninitiated, however, the shapes convey nothing, and I have known cases of Europeans who have been many years in the country, who were quite unsuspicious of what 'that jew's-harp idol,' as they called it, was intended to represent. As the old priest at Chandeshwar said, 'Those who can afford it, put up a big Mahádeo; those who can't, put up these slabs.'

"In the view that these markings are nothing but a conventional rendering of the Mahádeo and yoni, I am further confirmed by what has recently been brought to my remembrance of the manner in which an Amin, or native surveyor, will indicate a Mahádeo temple on his plotting. I remember that the sign used to mark the position of such temples by the Amins in the Field Survey of the Chandá Revenue Settlement, in which district I was Settlement Officer, and where this form of worship is very common, almost exactly resembled the rock-sculpturings in Kamam. They are not unlike the form of the vestal lamp. Indeed, on the summit of a hill near Ránikhet, on the top of a pile of stones which did duty for a Mahádeo shrine, I found a small slab, bearing an almost exact resemblance to the well-known form of the classic lamp. In the hole into which oil is poured, a small upright Mahádeo is placed."

Some years ago, Mr. J. H. Rivett-Carnac, of the Bengal Civil Service, in communicating his views respecting the cup and ring markings in the rocks of India, expressed himself rather strongly to the effect that the concentric circles and certain curious markings of what some have called the "jew's-harp type," so common in Europe, are traces of phallic worship, carried there by tribes whose hosts descended into India, pushed forward into the remotest corners of Europe, and, as their traces now seem to suggest, found their way on to the American continent also.

"Whether these markings really ever were intended to represent the phallus and the yoni, must always remain a matter of opinion, but," says Mr. R. Carnac, "I have no reason to be dissatisfied with the reception with which this, to many somewhat unpleasant, theory has been met in some of the antiquarian societies of Europe.

"No one who compares the stone yonis of Benares with the rock markings of Northumberland and Argyleshire will deny that there is an extraordinary resemblance between the conventional symbol of Siva worship of to-day and the ancient markings on the rocks, menhirs, and cromlechs of Northumberland, of Scotland, of Ireland, of Brittany, of Scandinavia, and other parts of Europe.

"A further examination of the forms of the cromlechs and tumuli and menhirs will suggest that the tumuli themselves were intended to indicate the symbols of the Mahádeo and yoni, conceived in no obscene sense, but as representing regeneration, the new life, 'life out of death, life everlasting,' which those buried in the tumuli, facing towards the sun in its meridian, were expected to enjoy in the hereafter. Professor Stephens, the well-known Scandinavian antiquary, speaks of these symbols as follows:—'There can be no doubt that it is to India we are to look for the solution of many of our difficult archæological questions.'

'But especially interesting is your paper on the Ancient Rock Sculpturings. I believe that you are quite right in your views. Nay, I go further. I think that the northern bulb-stones are explained by the same combination. I therefore send you by this post a copy of the part for 1874 of the Swedish Archæological Journal, containing Baron Herculius' excellent dissertation on these objects. . . . . You can examine the many excellent woodcuts. I look upon these things as late conventionalised abridgments of the linga and yoni, life out of death, life everlasting,—thus a fitting ornament for the graves of the departed.

'In the same way, the hitherto not understood small stones with one or two or three or four, etc., distinct cups cut in them (vulgarly called chipping-stones, which they never were or could be), I regard as the same thing for domestic worship, house altars, the family Penates.'"

Mr. Carnac remarks that many who indignantly repudiate the idea of the prevalence of phallic worship among our remote ancestors, hold that these symbols represent the snake or the sun. But admitting this, may not the snake, after all, have been but a symbol of the phallus? And the sun, the invigorating power of nature, has ever, I believe, been considered to represent the same idea, not necessarily obscene, but the great mystery of nature, the life transmitted from generation to generation, or as Professor Stephen puts it, "life out of death, life everlasting." The same idea, in fact, which, apart from any obscene conception, causes the rude Mahádeo and yoni to be worshipped daily by hundreds of thousands of Hindus.

Mr. Carnac afterwards came across the following by Tod in the Asiatic Researches:—"The Suroi were in fact the Sauras, inhabiting the peninsula of Saurashtra, the Saurastrene and Syrastrene already quoted from the Periplus, and the kingdom immediately adjoining that of Tessarioustus, to the eastward. That the Συροι of Saurashtra and the Syrians of Asia Minor had the same origin appears from the worship of Surya, or the Sun. I have little doubt we have more than one 'city of the sun' in this tract; indeed, the only temples of the sun I have met with in India are in Saurashtra. The temple raised to Bal, in Tadmor in the Desert, by Solomon, where he worshipped 'Bál and Ashtoreth, the strange gods of the Sidonians,' was the Bál-nat'h, or Great God of the Hindus, the Vivifier, the Sun : and the Pillar erected to him, 'in every grove, and on every high hill;' the Lingam or Phallus, the emblem of Bál; Bál-nath, Bāl-cesari, or as Bál-Iswara, the Osiris of the Egyptians; and as Nand-Iswara, their Serapis, or Lord of the Sacred Bull; Nanda, or Apis, 'the Calf of Egypt,' which the chosen people bowed to 'when their hearts were turned away from the Lord."

After the supreme Triad, which occupied the adytum of the temple at Hierapolis, came the personification of their various attributes and emanations, which are called after the names of the corresponding Grecian deities, and among which was an ancient statue of Apollo clothed and bearded, contrary to the usual mode of representing him. Lucian, *De Dea Syria*, says, "There is a statue of Apollo, not as was usual to make such, for all others represent Apollo young and in the attitude of running, but they have given him in this statue a beard. In another particular they have made an innovation in their Apollo : they have covered him with garments."

In the vestibule were two phalli of enormous magnitude, upon one of which a person resided during seven days twice in each year to communicate with the gods, and pray for the prosperity of Syria; and in the court were kept the sacred or symbolical animals, such as bulls, horses, lions, bears, eagles, etc. Lucian remarks upon this as follows :—" The two great phalli standing in the porch with the inscription on them : These phalli I, Bacchus, dedicated to my stepmother, Juno! The Greeks erect phalli to Bacchus, which are little men made out of wood, *bene nasatos*, and these are called neurospasta [moving by artificial muscles]. There is also on the right hand of the temple a little brazen man, whose symbol is enormously disproportionate. There is also in the temple the figure of a female, who is dressed in man's clothes. The priests are self-mutilated men, and they wear women's garments. The temple itself stands upon a hill, in the middle of a city (Hierapolis, the holy city, near Aleppo), and it is surrounded by a double wall. The porch of the temple fronteth the north, and it is two hundred yards in circumference ; within it are the two phalli before mentioned, each about a hundred and fifty yards high. To the top of one of these phallic pillars a man ascends twice during the year, and he there remains seven days at a time. The vulgar imagine that he converseth with the gods above and prayeth for the prosperity of all Syria, which prayers the gods hear, near at hand. He never sleeps during the seven days."

In the year 1845, Dr. Troost, of the University of Nashville, brought under the notice of the American Ethnological Society a number of ancient Indian remains in the shape of certain idols which had been discovered at various times in Tennessee. One was found when some new ground was ploughed, *i. e.*, ground which had never before been under cultivation, and was for the first time reclaimed from its primitive forest state. This idol was yet in its sanctuary, namely, in a large shell, the interior whorls and columella were all removed so that nothing but the external shell remained, which was opened in the front sufficiently to permit the image to enter.

Dr. Troost says:—" The utensils which I have found were all made of different kinds of stone, most of which may be found amongst the primitive rocks of North Carolina or Missouri States, and some even in Tennessee ; one, however, namely, an obsidian, must have been brought from South America. We know that this volcanic substance is found in several parts of the Andes, particularly in Quito, Popayan, at the volcanoes of Paracē and Sotora, the mountains Las Nobayas, and Mexico.

" These facts seem to place it beyond doubt that they came from southern regions. I have already observed that they were idolaters, and probably worshipped the phallus, as did some ancient nations—Egyptians, Phœnicians, and Greeks. I had the good fortune to obtain, during my rambles, several images which, no doubt, must have served for religious purposes ; they have all, at least such as were not too much mutilated, some similarity in their position,—they are all in a kneeling position and sitting on their heels, all without clothing. Some of them have their hands around their abdomen ; others have their hands on their knees. Two of these images, a male and a female, are the largest I have seen, being sixteen inches high ; they were found in Smith county, and are made of a kind of sandstone, and are of a rude sculpture. The male seems to be a rude imitation of an ancient Priapus ; he is more or less injured by the plough by which he was brought up, and which has broken a large *membrum generationis virile in erectione:* the marks of the ploughshare are yet visible, passing from that part over the face, breaking part of the chin and nose. The person who ploughed it up mentioned that it possessed this member, but he considered it too indecent to be preserved. It is not the only instance where this *pars genitalis* has been found. Doctor Ramsey, who has a fine collection of these antiquities, has two simulacra of this member: the one is carved out of stone similar to that of my images, and is of a rude construction, but he has one which is made of a kind of amphibolic rock, and perfectly resembling the natural object. The latter, if I remember right, is about twelve inches in length, the other four inches longer. The one made of amphibolic rock must have taken a long while to make, it being a very hard and tough rock on which steel can make no impression; it must have been ground down with a substance of the hardness of emery, nevertheless it is perfectly smooth, having the fat or greasy lustre characterising these rocks. It is not probable that they would have spent so long a time on an object merely to satisfy some voluptuous propensities or whims; they must have served some more serious purposes, and it is very probable that they held them in the same veneration as the Greeks, who consecrated the organ of generation in their mysteries. The phallus and steis were exposed in the sanctuaries of Eleusis. The Egyptians had consecrated the phallus in the mysteries of Osiris and Isis; and, if I am not mistaken, Father Kircher mentions, on the authority of Cortes, that this worship was established in America.

"I have another image made of stone similar to that of the two above mentioned, but it is only nine inches high: it much resembles the first. It was found in the Sequat-chy valley.

"A fifth, which was found in Sumner county, differs much in figure and in the materials of which it is made from the above mentioned images. It is made of clay mixed with pounded shells, and, judging from its red colour, it must have undergone a certain amount of heating; it is entirely hollow, and has a hole on the back of the head through which the finger was introduced to mould the clay in its present form. It is a female.

"The image in its sanctuary, mentioned above, is made of the same materials as the preceding, which it resembles except in its features. It was found in the Sequat-chy valley by ploughing. It was in its chapel as observed above."

During his investigation, Professor Joseph Jones obtained from the tumuli and valleys of Tennessee several interesting idols, both of stone and of clay mixed with pounded shells. It may be stated that images of this archaic type have been found also in Kentucky, Virginia, South and North Carolina, Louisiana, Alabama, and Florida. The worship of the Priapus probably obtained among some of the southern Indian nations. In the collection of Dr. Troost were many carefully-carved representations in stone of the male organ of generation. They were found principally within the present limits of the State of Tennessee. But two objects of this sort appear to have been noted among the relics of the Georgia tribes, and they were about twelve inches long, made of slate. In some parts of Alabama, and in Mississippi, similar objects have been exhumed from grave mounds.*

"Phallic emblems abounded at Heliopolis, in Syria. Not having any knowledge of their existence at Heliopolis, in Egypt, I took means to ascertain it from a brother physician who had recently visited the country. The following is his reply to my query:—'I am very sorry that I am not enough of an antiquarian to give you much information on the subject you are interested in. I was in Egypt last winter (1865-6), and there certainly are numerous figures of gods and kings on the walls of the temple at Thebes, depicted with the penis erect. The great temple at Karnak is, in particular, full of such figures, and the temple of Danclesa likewise, though that is of a much later date, and built merely in imitation of the old Egyptian art. I remember the scene of a king (Rameses II.) returning in triumph with captives, many of whom are undergoing the operation of castration, and in the corner of the scene are numerous heaps of the complete genitals which have been cut off, many hundreds in all, I should think. This is on the walls of Medinet Haboo, at Thebes.'

"This letter," says Dr. Inman, "is very interesting, for it shows (1) how largely the idea of virility was interwoven with religion; (2) how completely English Egyptologists have suppressed a portion of the facts in the histories which they have given to the world; (3) because it tells us of the antiquity of the practice which still obtains among the negroes of Northern Africa, of mutilating entirely every male captive and slain enemy (compare 2 Kings xx. 18, and Isa. xxxix. 7, also 1 Sam. xviii. 25-27). In Assyria and Palestine conquerors counted the heads of the slain, which were piled in heaps before them. The learned

* Jones. *Antiquities of Southern Indians.*

Egyptians were content with a less bulky emblem. A man when beheaded is useless; if only emasculated he is of value as a slave. The Asiatic gratified a temporary revenge; the African had an enduring triumph."

M. Rivett-Carnac says:—" No one who has been in this country (India), and who has noticed the monolith Mahádeos of the Western Gháts of the Himalayas and other parts of India, can fail to be struck with the resemblance that the menhirs of Carnac in Brittany and its neighbourhood bear to the Siva emblems of India. I visited these remarkable remains when at home last year, and was quite taken aback by their resemblance to well-known Indian types. The monoliths of Scotland, covered with what I believe to be Mahádeo symbols, are of the same class. Added to this, in the recesses of the Pyrenees, the people whose language suggests their descent from the tribes who erected the tumuli and menhirs, not only in this neighbourhood but also in other parts of Europe, still preserve traditions connected with these monoliths, and have actually retained some traces of what I call Siva worship."

MM. Edouard Piette and Julien Sacaze prepared a paper some twelve years ago, which was read at the Society of Antiquaries of France in which they shewed that they had actually discovered the remains of phallic worship still existing among the people of the Pyrenees, the existence of which, in Scandinavia, in days gone by, had already been brought to the notice of the Society by Dr. Rajendralala Mitra. These archæologists have established the fact that to this day the menhir is still reverenced in the Pyrenees as the phallus.

Captain Richard J. Burton, late Commissioner to Dahome, says, in his " Notes connected with the Dahoman :"—" Amongst all barbarians whose primal want is progeny, we observe a greater or less development of the phallic worship. In Dahome it is uncomfortably prominent; every street from Whydah to the capital is adorned with the symbol, and the old ones are not removed. The Dahoman Priapus is a clay figure of any size between a giant and the pigmy, crouched upon the ground as if contemplating its own attributes. The head is sometimes a wooden block rudely carved, more often dried mud, and the eyes and teeth are supplied by cowries. A huge penis, rudely carved as the Japanese articles which I have lately been permitted to inspect, projects horizontally from the middle. I could have carried off a donkey's load had I been aware of the rapidly rising value of phallic specimens amongst the collectors of Europe. The Tree of Life is anointed with palm oil, which drips into a pot or a shard placed below it, and the would-be mother of children prays that the great god Legba will make her fertile. Female Legbas are rare—about one to a dozen males. They are, if possible, more hideous and gorilla-like than those of the other sex; their breasts resemble the halves of German sausages, and the external labia, which are adored by being anointed with oil, are painfully developed. There is another phallic god, named 'Bo,' the guardian of warriors and the protector of markets."

CHAPTER V.

Wide Prevalence of Phallic Worship.

THE Rev. J. Roberts, in a paper read before the Royal Asiatic Society in 1832, said :—"That the gods adored by the Israelites, taken from the Assyrian and other nations, are still served by the Hindus (though generally under other names) I cannot doubt, and the object of the following observation is to identify some of the leading deities. It has been well observed :— ' Whoever were the first planters of India, it could not have been planted till long after Persia and Elam had been sufficiently cultivated, and a considerable number of ages after Assyria and the countries adjoining Ararat had been planted. This is so apparent, both from Scripture and the nature of things, that it will not admit of a dispute.' Is it not reasonable to suppose that Noah and his family would remain for many years at no very great distance from the spot where they first settled? Who built the splendid cities of Babel and Nineveh? did not Ashur, and probably the other sons of Noah? Who were the first to study astrology as a guide to find out the good or evil supposed to be produced by the heavenly bodies? Who were the first to propitiate them in reference to their salutary or malignant influences on the destinies of men? Does not the mind immediately revert to the builders and occupiers of Babylon; to their dispersion over the earth; and the consequent carrying away of their superstitions, though then veiled in different languages? If then, 'India was peopled after Persia and Elam, and many years after Assyria,' from whom did she receive her leading deities and theological institutes? Is it not natural to suppose from one of the above? And from whom so likely as the Assyrians?

"The Jews worshipped the Assyrian deity Succoth-Benoth, under the name of Ashtoreth or Astarte; and it is said that this god or goddess was both masculine and feminine. The Siva of India is both male and female; his right side being of the former and his left of the latter sex; and his wife assumed both appearances as circumstances might require.

"The Babylonians called Succoth-Benoth, Mylitta, signifying mother. The wife of Siva, and she only as far as I know, is called Mátá or Mother.

"Amongst the Assyrians, the daughters or women once in their lives had to make a sacrifice of virtue to that goddess Succoth-Benoth. And Lempriere says of her:—"A surname of Venus among the Assyrians, in whose temples all the women were obliged to prostitute themselves to strangers. The wife of Siva, amongst many other names, is called Váli or Báli, under which appellation she assumed the form of a girl of twelve years of age. And in Madura, Balane, and other places, beautiful virgins used to go to the temple once in their lives to offer themselves in honour of the goddess. The story was that a god had converse with them. In all the temples of Siva and his consort (where it could be afforded) women were kept to dance and sing before the idols.

"Amongst the Assyrians and others, the votaries of the above-named goddess worshipped sometimes in the dress of men, and at other times in that of women. The dancing girls of many of the temples on the continent of India, at the feast called Manampu, do the same thing. When the god and goddess go out to hunt, they are equipped and mounted as men; and at the conclusion of the great feast of Siva they assume the dress of Pandárams, and thus go forth from house to house to ask alms.

"The Babylonian or Assyrian goddess was drawn or supported by lions. The wife of Siva, under the name of Bhadra-Kali, has the same animal appropriated to her use.

"Succoth-Benoth, the same with the Syrian goddess, the same as Astarte of the Phœnicians and the Decerts of Ascalon. The worship paid to this goddess came originally from Assyria and Babylonia. Astarte is always joined with Baal; and is called a god in Scripture, having no particular word for expressing a goddess. Lucian thinks Astarte to be the moon.

"The wife of Siva, under the name of Sacti, placed a representation of the cresent moon on the head of her husband under the following circumstances. When once engaged in armourous sports he by accident broke her arm-ring, which she immediately tied on his dishevelled lock of hair as the crescent moon. He, however, having laughed at her, she turned away her face and changed the crescent into full moon. The crescent is common to both, and is assumed as circumstances may require.

"Shach, or Saca, another god or goddess, partly the same with Mylitta, the Syrian goddess.

"The wife of Siva is also known under the name of Satti; but in Sanscrit, Sakti.

"The festival of Saca was held for five days every year; during which time servants commanded their masters, and wore a kind of royal garment called Zogani. *

"The festival of the wife of Siva continued nine days, or rather nights, and was called Nava Rattiri, *i.e.*, nine nights: three of these, however, were for Sarasvati, and the other six for Sakti. On this occasion, those who had not been accustomed to eat flesh, or drink intoxicating liquors, did so freely. All restraints were now thrown off, and scenes of the most sickening kind wound up the ceremonies. No young female of respectable character dared shew herself in public. Servants assumed the airs and practices of their masters; school-boys dressed in gay apparel, went from house to house to dance and sing songs in honour of Sakti: gambling, fighting of cocks and of rams, with other rude and ludicrous performances, filled up this indecent festival.

"Salambo, a goddess; the same as Astarte; eternally roaming up and down a mountain. It is rather striking that the wife of Siva is also known by the name of Silambu, and that this name also signifies a mountain. Another of her names is Parvati, meaning she who was born on a mountain. She is also called daughter of the mountain; and sometimes the mountain nymph, who captivated Siva from a course of ascetic austerities.

"The Babylonians and Assyrians worshipped what by the Greeks and Romans was termed phallus or Priapus. The priaps were three hundred fathoms, or three hundred cubits high; and by whom the priaps were executed there is much dispute.

"The Egyptians most probably meant the sun and moon. Some suppose Osiris to signify the efficient cause of things; and Isis, matter. Osiris was represented in a human form, in a posture not very decent, signifying his generative and nutritive faculty. This living image was the bull. The image of Isis, usually in the form of a woman with a cow's horns on her head.

"Calmet regards Astarte, or Astaroth, as the Isis of Egypt. The word is often plural, Astaroth. Sometimes aserah, the grove; aseroth, or aserim, woods; groves were her temples: in

* *Universal History.*

groves consecrated to her, such lasciviousness was committed as rendered her worship infamous. She was also called the queen of heaven, and sometimes her worship is described by that of the *host of heaven*. She is almost always joined with Baal, and is called gods.

"Temples of the moon generally accompanied those of the sun; and while bloody sacrifices, or human victims were offered to Baal, bread, liquors and perfumes were presented to Astarte; tables were prepared for her on the flat terrace roofs of houses, near gates, in porches, and at cross-ways, on the first day of every month, which the Greeks called Hecate's supper.

"St. Jerome, in several places, translates the name Astarte by Priapus, as if to denote the licentiousness committed in her groves. The eastern people in many places worshipped the moon as a god: represented its figure with a beard, and in armour. The statue in the temple of Heliopolis, in Syria, was that of a woman, clothed like a man, says Pliny. Solomon, seduced by his foreign wives, introduced the worship of Astarte into Israel, but Jezebel, daughter of the king of Tyre, and wife of Ahab, principally established her worship.

"St. Austin assures us that the Africans maintained Astarte to be Juno. Herodian says the Carthaginians call the heavenly goddess the moon, Astroarche. The Phœnicians asserted confidently, says Cicero, that their Astarte was the Syrian Venus, born at Tyre, and wife to Adonis; very different from the Venus of Cyprus. Lucian, who wrote particularly concerning the goddess of Syria (Astarte), says expressly that she is the moon, and no other, and it is indubitable that this luminary was worshipped under different names in the East.

"The manner of representing Astarte on medals is not always the same. Sometimes she is in a long habit, at other times in a short habit; sometimes holding a long stick, with a cross on its top; sometimes she has a crown of rays; sometimes she is crowned with battlements, or by a victory.

"In regard to the indecent object," wrote Mr. Roberts, "alluded to as being worshipped by the Assyrians, it is well known that the Hindus do the same thing. The *lingam* (priapus) in the Hindu temple of Sedambarem is supposed to have sprung from the earth of itself, and its foundation is believed to be in the lower world.

"Buckingham says, in his Travels in Mesopotamia, vol. II., page 406, of some antiquities he saw taken from the ruins of

Babylon :—'The larger antiques comprehended a figure in brass, embracing a large lingham between its knees, precisely in the style of the Hindu representation of that emblem.' He mentions also in another place, 'the Indian figure of a man, with a painted bonnet and beard, embracing the lingham.'

"In regard to Osiris, it is more than probable that he in his posture, generative and nutritive faculties, was the same as the Siva of the Hindus. The bull was sacred to the former, and also to the latter. Isis being represented with cow's horns, finds a parallel in Siva or his wife, with the crescent moon fixed on her head.

"Whether we look at the corresponding traits of character in Moloch and Kali; in Baal-Peor and the Chiun of Anos; at the mutual assumption of either sex by Siva and his partner; at the term mother being applied to the latter, and also to the Succoth-Benoth of the Assyrian, Phœnician, and other nations; at the cow's horn (so called) of Assyria, and the crescent of India; at the young virgins who made a sacrifice of chastity to the Succoth-Benoth of antiquity, and to the consort of the oriental Siva; at the use made of the regular female votaries of both systems; at their mutual assumption, on certain occasions, of the male attire; at the lion, as belonging to the goddess of Assyria, and also to her of India; to the festival of Shach or Saca, and that of Satti or Sakti, in regard to the lascivious way in which it was conducted, and the peculiar garments worn on that occasion; at the term Salambo being the name of the one goddess, and also of the other; at its true meaning in reference to a mountain where they mutually dwelt; at the Baal-Peor of Assyria, the Osiris of Egypt, the Phallus of the Greeks, the Priapus of the Romans, and the Lingam of the Hindus (worshipped now in the temples of the East), we see some of the most striking coincidences, which never could have been the result of anything but the identity of their origin."

"The Chaldees believed in a celestial virgin, who had purity of body, loveliness of person, and tenderness of affection, and who was one to whom the erring sinner could appeal with more chance of success than to a stern father. She was pourtrayed as a mother with a child in her arms, and every attribute ascribed to her showed that she was supposed to be as fond as any earthly female ever was. Her full womb was thought to be teeming with blessings, and everything which could remind a votary of a lovely woman was adopted into her worship.

"The worship of the woman by man naturally led to developments which our comparatively sensitive natures shun, as being opposed to all religious feeling. But amongst a people whose language was without the gloss of modern politeness—whose priests both spoke and wrote without the least disguise, and whose God, through His prophets and law-givers, promised abundance of offspring and increase in flocks and herds as one of the greatest blessings he had to bestow—we can readily believe that what we call 'obscenities' might be regarded as sacred homage or divine emblems.

"In India, at the present time, both the thoughts and conversation of the lords of the soil turn, unpleasantly to us, upon the power possessed by each to propagate his race, and European doctors are more frequently consulted for the increase or restoration of this power than for any other cause.

"Not only does the man think thus, but the female has her thoughts directed to the same channel, and there has been a special hell invented by Hindoo priests for childless females. It is curious to see at India's antipodes a similar idea started amongst the offshoots of a Christian community, but so it is, and Mormon women join themselves in numbers to the man from the belief that without union with him they cannot attain to celestial glory.

"The Bible student will remember the plaintive entreaty of Rachel, 'Give me children, or else I die' (Gen. xxx. 1); the earnest prayer of Hannah, and the spiteful persecution of Peninnah (1 Sam. 1); and he will recall the longing for offspring which induced Abraham to consort himself with a black (Egyptian) slave girl, and how complaisant his wife was in delegating for a time her rights (Gen. 14).

"In Deuteronomy xxviii., we find 'the fruit of the body' promised as one of the special blessings for obedience to the law, and in Psalm cxxvii. 3, we are distinctly told that 'children are an heritage of the Lord, and the fruit of the womb is his reward.'

"If abundance of offspring was promised as a blessing, it is clear to the physiologist that the pledge implies abundance of vigour in the man as well as in the woman. With a husband incompetent, no wife could be fruitful. The condition, therefore, of the necessary organs was intimately associated with the divine blessing or curse, and the impotent man then would as naturally go to the priest to be cured of his infirmity as we of to-day go to the physician. We have evidence that masses have been said, saints invoked, and offerings presented, for curing the debility we refer to, in a church in Christianised Italy during the last hundred years, and in France so late as the sixteenth century,—evident relics of more ancient times." *

In further elucidation of the above as set forth by R. P. Knight in his "Worship of Priapus," Dr. Inman says :—" Whenever a votary applied to the oracle for help to enable him to perform his duties as a prospective father, or to remove that frigidity which he had been taught to believe was a proof of Divine displeasure, or an evidence of his being bewitched by a malignant demon, it is natural to believe that the priest would act partly as a man of sense, though chiefly as a minister of God. He would go through, or enjoin attendance on, certain religious ceremonies—would sell a charmed image, or use some holy oil, invented and blessed by a god or saint, as was done at Isernia—or he would do something else.

"We can readily see, then, how some sacred rites might be intentionally provocative of sexual ideas; how desirable it might have been for hierarchs to compose love philtres or satyrion, and to understand the influence over the sexual powers possessed by various kinds of aliment; and we can also understand how certain gods would be invented whose images should act as amulets, and who, like special Roman saints, would take charge of this particular part of the body.

"Even after 'the Reformation,' France introduced Saint Foutin into the Christian calendar, to whom offerings were made by the faithful who found themselves unable to procure the blessing of fertility;—they are thus described :—

"'Temoin Saint Foutin de Varailles, en Provence, auquel sont dediées les parties honteuses de l'un et de l'autre sexe, formées en cire ; le plancher de la chapelle en est fort garni, et quand le vent les fait entrebattre, cela débauche un peu les devotions à l'honneur de ce saint." (La Confession de Sancy, vol. v., Journal de Henri III., by Pierre de l'Etoile, ed. Duchat, pp. 383, 391).

"This and other saints were worshipped for similar purposes, as St. Guerlichon, or Greluchon, and St. Entropius, at Orange, Porigny, Vendre, Auxerre, Puy en Velay, in the convent in Girouet, and at Bourg Dieu ; St. Gilles in Brittany ; St. René in Anjou ; St. Regnaud in Burgundy ; St. Arnaud and St. Guig-

---

* R. P. Knight on Priapus cited by Inman.

nolé, near Brest and in Berri. The worship of many of these was in full practice in the last century." (Two Essays on the Worship of Priapus; London, 1865 : privately printed.)

"If," says Dr. Inman, "with all the vaunted enlightenment of Christian Europe, there are several canonised mortals whose special care, in the heaven to which they have been promoted by men on earth, is to help unfortunates who require their aid 'pour les parties honteuses,' we cannot wonder that sexual saints should be found amongst the heathen races of Asia; nor can we refuse credence to the idea that the act of propagation was sometimes the end of certain forms of worship, which were specially adapted to bring about that act.

"As a physician, I know how much intense misery is felt by those men who, from any cause, are unable to do their part in multiplying their race. I can readily understand that a cure of impuissance would raise to the highest pitch, in the mind of an enquiring devotee, his estimate of the saint who wrought it, and I do not see why masses should not be said to St. Greluchon, for raising the courage of the living, as much as to St. Denis (or Dionysius), for the consolation of the dead. At any rate, the Chaldees used some of their gods, or divinities, for comparatively a holy worship, and for a cult as peculiar as that paid to the modern Priapus, St. Foutin."

The Universal History tells us that "At Hierapolis, or the Holy City, or Magog, as the Syrians themselves are said to have called it, in the province of Cyrrhestica, stood the temple of the great Syrian goddess. It is impossible to say whom they meant by this Syrian goddess, but we find a story in Justin, which we suppose to be borrowed from Nicolas of Damascus, that a king from whom Damascus derived its name had a queen called Arathis, whose sepulchre was religiously frequented by the Syrians, who esteemed her as their principal deity, and this queen, according to our author's account, was older than Abraham, whom he reckons among the kings of Damascus.

"The temple was upon an eminence in the midst of the city, surrounded by a double enclosure or two walls, the one old, the other new. At the north side it had a court, or porch, before it of about five or six hundred feet in circumference, where stood the priaps of three hundred fathoms, or three hundred cubits high; for we find both these measures, but both exceedingly too large, to our apprehension. These obscene images, or rather columns, were but slender, as we shall shew hereafter, but by whom, or to whom, they were erected was the subject of much fable. The front of the temple itself stood last, and before it was a tower raised upon a terrace about twelve feet high, which was no sooner mounted than the temple appeared. It was built after the manner of the Ionian temples; the porch of it was adorned with golden doors, nay, the whole temple glittered with gold, and particularly the roof; the air about it was enchanting, nothing inferior to the sweetest of Arabia, and so strongly perfumed the garments of all that visited it that they were scented for a considerable time."

A considerable number of festivals were observed by the people of this place, and many singular customs indulged in. Of priests there were several sorts, each assigned to particular tasks. Some killed the sacrifices, some bore the drink offerings, some carried fire, and some waited at the altar, and of these above three hundred, in white habits and with caps or bonnets on their heads, attended the sacrifices. And besides them, there were other consecrated orders: as minstrels, skilful in the touch of several instruments; Galli, or eunuch priests, and mad and frantic women. The office of high priest was annual; he wore purple, and a golden mitre. There were other sorts of holy persons of various nations who held the Syrian goddess in veneration, whose business it was to instruct their countrymen, who from time to time resorted thither in pilgrimage, in the rules and customs of the holy city. They were called masters or instructors.

It is remarkable of their eunuch priests that they were emasculated by the voluntary operation of their own hands. How this unnatural custom came to prevail is accounted for by the following story. Stratonice, who built this temple, having been admonished by the goddess to undertake the work, took no heed to her purpose till she forcibly brought her to obedience by a violent procedure against her, and the king, her husband, consented to let her go and take in hand the building, but committed the care of her to a beautiful youth named Combabus. This Combabus, no way fond of his commission, but dreading the consequences of being so much alone with the beauteous queen, destroyed his sex, and left the ruins of it, carefully embalmed and sealed up, with the king. He departed with the queen Stratonice, was after some time solicited by her, and convinced her of his inability. But, nevertheless, it came to pass, either through malice or envy, that

Combabus was clearly convicted of adultery, infidelity, and impiety to the goddess. As he was being led to execution he called out for the treasure he had left with the king, which being produced, his intended punishment was converted into the most tender embraces in the arms of his prince. The king now raised him to the highest degree of riches and honour, and he was famed for the wisest and happiest man living. Combabus desired leave to finish the temple, which being granted, he passed the remainder of his life there. And there stood his statue in brass, the work of Hermolaus the Rhodian. And because it was reported that some of his dearest companions resolved to undergo his calamity for his sake, or that Juno inspired several with a rage of unmanning themselves, that he might not be single in his misfortune, many mad zealots, either in honour of Combabus or to please Juno, performed the hated operation on themselves every year in the temple. Furthermore, these Galli, or devoted eunuchs, took on them the habit and offices of women, because Combabus had been fallen in love with by a strange woman, who knew not the violence he had done to his sex, which to prevent for the future, he put on the outward appearance of a woman.

Of all the holy days of this people the *Great Burning*, as it may be called, was the principal. Upon this occasion there came people from all parts to assist at the great sacrifice, and all the other religious duties of the season. This festival was of some days' continuance, and at particular times, while it lasted, the whole multitude was drawn into the temple, while the priests stood without, some of them mangling their bodies, some thumping violently against each other, while others beat upon tabrets or drums, and sounded musical instruments, and sang and prophesied. And now it was that, amidst all this uproar, the frenzy of castrating themselves seized on many in the temple, who, crying out with a loud voice and drawing their swords, performed the operation and devoted themselves to the goddess.

The religious customs, and laws, and traditions of this place were as extraordinary as any to be found in any part of the world. Twice a year a man went up to the top of one of the priaps, and there remained seven days. He got up in this manner :—He surrounded the priap and himself with a chain, and ascended by the help of that and certain pegs or pins, which stuck out of the sides of the priap for that purpose, lifting the chain up after him at every step he mounted. It has been said that those who had seen men climb up the palm-trees in Egypt and Arabia might readily understand him, from which it has been gathered that these phalli or priaps, so monstrously high, were so slender that a man might grasp them. In whatever way we reckon the heights as given by different authorities, they would be about the height of the monument erected to commemorate the Fire of London at the foot of London Bridge. "How so slender a body could be reared to such a height," remarks an old writer, "we leave to those who are better skilled in such matters than ourselves."

When the man had got up, he let down a chain, wherewith he drew up all such things as he required and therewith made a seat or kind of nest for himself. It was given out that during the seven days he had a more intimate acquaintance with the great goddess, and also that this was done in memory of Deucalion's flood, when the men climbed up the mountains and trees to save themselves from perishing. During the seven days it is reported that the man never slept, and that if at any time he happened to doze, a sacred scorpion went up and awakened him; it is said that the fear of falling kept him awake.

More than half-a-century ago, a writer on Freemasonry in discussing the question of the cause of the original dispersion of primitive nations, promulgated a number of ideas respecting towers and pyramids which had a very close connection with the subject now under our consideration. He asked what was really intended, mystically figured and represented under the colossal and other monuments and sacred edifices of antiquity, specially designating as the head and type of all succeeding edifices of like character the Tower of Chaldea and the Great Pyramids of Egypt. The first of these was erected not long after the foundation of the Chaldean monarchy by Nimrod, the son of Cush, 2221 B.C. The second were erected probably not long after the foundation of the Egyptian monarchy by Misraim, the son of Ham, 2188 B.C., Babylon and Memphis being among the first cities built after the Flood. And when the totally different forms of these immense national edifices are considered, the enquiring mind can scarcely fail to seek for the causes which decided their ancient architects to employ so gigantic a mass of materials, in one or the other of these definite forms, above that of every other which might have been selected, and "It will scarcely be denied," said the writer in question, "that the forms respectively of these stupendous monuments (which were only the original archetypes of innumerable others which have been subsequently constructed) must unavoidably be considered as having been adopted as the carrying out of some paramount idea or intention on the part of their primæval founders.

"There is cause to believe," this gentleman thinks, "that in the erection of the Chaldean tower, the principles of true Masonry were at first abided by, but subsequently, the corruption of human nature urging men to overthrow a spiritual worship which absolutely required purity and holiness, they sought to establish a system which virtually inculcated the worship of the creature more than the Creator, and furnished a pretext for the practice of unrestrained licentiousness as part and parcel of religious rites. Such was the ancient worship of the Lingam—a worship which we read of as recognised and established throughout all antiquity." "Such," he says, "was the object really worshipped under its colossal representative in the Chaldean Tower, of whose notorious existence traditions even in the most remote nations, almost universally exist, and of whose actual signification many weighty proofs have been collected by the late Mr. O'Brien, to which I might certainly add others equally numerous and irrefragable."

The worship of the Lingam, then, of which the pillar tower was a gigantic figure, involved and signified the worship of the male principle of the universe; this worship, though afterwards perverted, originally intended the worship of the true and only God, in accordance with which assertion we find that one interpretation of the word Jehovah undoubtedly signifies the Universal Male. In India, where undeniable proofs have been found of the existence at one period of true Masonry, this signification is found to be involved in the names of the principal deities. Thus, according to Sir W. Jones, Bhagavat signifies the first male, and Naravan, or moving on the waters. The Spirit of God is often likewise denominated the primæval male. The deity described in the fourth Veda as Mahapurusha also signifies the Great Male. Accordingly, we find that temples in honour of this Universal Male Power took definite form, and were always erected, in the figure of its representative, the Lingam; that is to say, in the form of a tower or column. Almost innumerable examples of such-like edifices abound in ancient countries, where this worship was either primitive or introduced at later periods, and fully illustrate these facts.

Wilford remarks that the phallus was publicly worshipped by the name of Baleswara Linga on the banks of the Euphrates. The cubic room in the cave of Elephanta likewise contains the Lingam, as does also the pagoda of stone at Miherbaliporam, or *City of the Great Baal.* Sir W. Jones observes, "Columns were erected, perhaps as gnomons, others probably to represent the phallus of Iswara." Enough has here been cited, without doubt, to dispose both the learned and the unlearned to consider that the true signification of the pillar and tower was in reality such as has here been stated.

In many parts of the Bible we find the pillar to have been undoubtedly a sacred emblem; as in Isaiah xix. 19, "In that day shall there be an altar to Jehovah, in the midst of the land of Egypt, and a pillar at the border thereof, to Jehovah, and it shall be for a sign, and a witness to the Lord." And this was the especial form in which God Himself is described as appearing, when He dwelt in the pillar that went before His chosen people, as recorded by Moses.

When, however, pillars were set up to receive the profane rites of idolatrous worship, we find them noticed in Scripture as an abomination, in like manner as their great Babylonian archetype; which, being obnoxious to the wrath of God, as such, was destroyed by fire from heaven, as its blasted and vitrified ruins still remain incontrovertibly to attest. To this peculiar idolatry Scripture refers in the following passages: Deut. xii. 3, "Ye shall overthrow their altars, and break their *pillars,*" &c.; Leuit. xxvi. 1, "Ye shall make ye no idols, nor graven image, neither rear ye up a standing image" (Heb. pillar); 1 Kings xiv. 23, "For they also built them high places and images (Heb. *standing images*), on every high hill"; Ezek. xvi. 17, 24, 25, "Thou madest to thyself images of men (Heb. of a male), and didst commit," &c. Also Jer. xi 13, "According to the number of the streets of Jerusalem have ye set up altars to that shameful thing," &c. The same is also alluded to in the striking history recorded in Judges vi. 32, "Therefore he called the idol Jerubbaal (or Jerubbesheth, Heb.), *i.e.*, let the shameful thing plead." And a final warning was given to the Israelites by Moses as recorded in Deut. iv. 15, 16, "Take ye therefore good heed unto yourselves, for ye saw no manner of similitude on the day that Jehovah spake unto you in Horeb, out of the midst of the fire: lest ye corrupt yourselves, and make you a graven image, the similitude of any figure, the likeness of male or female."

As the tower was sacred to the male power of the universe, so likewise was the pyramid, triangle, or cone adopted by the votaries of an opposite worship as the real and consecrated emblem and representation of that procreative female energy in which (considering it as the true and vital conceptive power of nature), according to them, resided absolutely and solely, the underived principle of life; which female power they chose *alone* to deify, and, like their opponents, consecrated their unhallowed worship by the most profane and licentious rites.

Thus, the great pyramids were at Memphis the colossal monuments of a separate worship, with all its concomitant mysteries, and, like the Tower of Babel, in both symbolical edifices the threefold objects of astronomy, astrology, and religion were indissolubly involved and united.

Baron Humboldt observes, in his Researches (in total ignorance, however, of this theory), "In every part of the globe, on the ridge of the Cordilleras as well as in the Isle of Samothrace, in the Ægean Sea, fragments of primitive languages are preserved in religious rites."

Let us, in accordance with this observation, examine the ancient Sanscrit word YU, or YONI, which we shall recognise in the religious vocabulary of every nation where pyramidal edifices prove them to have been addicted to the schismatic worship of the Pish-de-Danaan sect. In the third volume of the Asiatic Researches Wilford says:—"Yavana is a regular participial form of the root *yu*, to *mix;* so that *yavana*, like *misra*, might have signified no more than a *mingled* people: but since *yoni*, or the *female nature*, is also derived from the same root, many Pandits insist that the Yavanas were so named from their obstinate assertion of a superior influence in the female over the linga, or male nature, in producing a perfect offspring." Sir William Jones expressly states that the meaning of yoni, or bhaga, is undoubtedly the female womb, and in his plate of the Hindu lunar mansions (in the article on the antiquity of the Indian Zodiac) this constellation of the yoni is figured as three stars, inclosed by the Hindu draughtsman in a representation of that object; which in his figure is made to resemble an inverted pyramid, or truncated cone. Venus Genetrix is sometimes represented in the form of a conical marble, "for the reason of which figure," says Tacitus, "we are left in the dark," "but," adds Sir William Jones, "the reason appears too clearly in the

temples and paintings of Hindustan, where it never seems to
have entered the heads of the legislators or people that anything
natural could be offensively indecent." Wilford mentions that,
according to Theodoret, Arnobius, and Clemens Alexandria, the
YONI of the Hindus was the sole object of veneration in the
mysteries of Eleusis.

For proofs of the high antiquity of this worship in China,
the discerning mind need only consult the following passage from
Lord Macartney's Travels, vol. I., Hager, Monument of Yu:
"In both Americas, it is a matter of inquiry what was the intention of the natives when they raised so many artificial pyramidal hills, several of which appear to have served neither as
tombs, nor watch towers, nor the base of a temple: a custom
established in Eastern Asia may throw some light on this
important question. About 2,000 years before our era, sacrifices
were offered in China to the Supreme Being, on four great mountains, called Four Yo. The sovereigns, finding it incovenient to
go thither in person, caused eminences representing these mountains to be created by the hands of men, near their habitations."
Thus we find that the pyrimidal hills on which sacrifice was
offered were designated by the very name of the object typified,
thus affording an invaluable clue to our present inquiries. (See
also Asiatic Researches, vol. III.) And the names of many of
their most considerable cities and provinces still bear witness to
religious or sacred etymology of their titles, and Yo-tcheon, Ya-ogan, in the province of Yunan, Yuengang, &c., are assuredly
derived from a primitively religious meaning, as the four sacred
Yo (or natural pyramids) themselves, alluded to by Lord Macartney.

The whole country of Mexico abounded in pyramids, and
Humboldt declares the basis of the Greek Cholula to have been
twice as broad as that of the Egyptian Cheops, though its height
is little more than that of Mycerinus. He states, also, that in
the American languages 137 roots have been recognised in the
languages of Asia and Europe, and in perfect accordance with
this theory of their religious ideas we find a Mexican province
named Yucatan; the Mexican Isis (Ish Ish, or female nature);
the wife of the sun is called Yu-becaygua-ya; and a chapel
called Yo-pico was built over the cavern that contained the skins
of the human sacrifices.

"It is extremely remarkable," says Humboldt, "that we
discover among the Mexican hieroglyphics absolutely nothing

which announces the symbol of the Generative Force, or the Lingam. M. Zoega has observed that the emblem of the phallus is likewise *never found in Egyptian works of great antiquity.* M. Larytes observes expressly that in India some sectarians have held this emblem in horror, " might we not suppose," adds he, " that there exists some exiled sect in the north-west of Asia who reject the worship of the Lingam, and of which one finds some traces amongst the American races ?"

Thus did these great writers, intuitively as it were, discern the existence of a separate object of worship in those countries which they allude to, although, as they never advert to the real nature of it, we may conclude their inferences to have been only obscurely conjectural.

The fact is that wherever this peculiar worship has ever flourished, traces are left behind, and relics remain which have always been found to have puzzled the learned antiquarian no less than the unlettered conjecturer.

" In a tumulus on Salisbury Plain," says Sir R. Hoare in his Tumuli Wiltunenses, " we found a cone of jet, likewise an amber cone. In another, we found an earthen cup of singular pattern, a cone of gold, &c.

That the mitre of Osiris, which, in fact, represented a *truncated cone,* had relation to this mysterious type, we imagine few will be inclined to controvert, and probably these remarks will receive corroboration from a close inspection of the extraordinary groups in the caves of Ellora (which are engraved in the sixth volume of the Asiatic Researches), where the mitre, placed on the head of Indra as well as on the head of Indranee, no less than the peculiar situation of the human skull and cross-bones, placed conspicuously on this last female personage, and occupying the natural position of the yoni itself, sufficiently evince what were the notions of those ancient architects concerning religious mysteries, and of the particular agency by which death, in the first instance, " passed upon all men."

A writer has said :—" In commenting on this particular branch of idolatry under discussion, we cannot but remark that there appears just reason to believe that this was the peculiar abomination into which the ten tribes of Israel lapsed, at their separation from Judah under Jeroboam ; of which opinion strong presumptive proof is offered, inasmuch as, from the account given by Herodotus (and cited by Josephus) of the invasion of the

Egyptian Shishak, under Rehoboam, it appears that, having conquered Jerusalem, and defiled the public buildings by carving in them the distinctive symbols of his own peculiar and national creed, that is to say (according to the same author), by defacing them with representations of that very symbol, the mysterious yoni, he returned to his own country without in any way molesting Samaria, the residence of the ten tribes, who, it needs not any great measure of sagacity to perceive, had doubtless embraced his religious views." *

* See *Freemason's Quarterly Review* for 1840.

CHAPTER VI.

Phallic Worship in the Middle Ages.

THE author of the treatise on the "Worship of the Generative Powers during the Middle Ages," in continuation of Payne Knight's book, says:—" Richard Payne Knight has written with great learning on the origin and history of the worship of Priapus among the ancients. This worship, which was but a part of that of the generative powers, appears to have been the most ancient of the superstitions of the human race, has prevailed more or less among all known peoples before the introduction of Christianity, and singularly enough, so deeply it seems to have been implanted in human nature that even the promulgation of the Gospel did not abolish it, for it continued to exist, accepted and often encouraged by the mediæval clergy. The occasion of Payne Knight's work was the discovery that this worship continued to prevail in his time in a very remarkable form, at Isernia, in the kingdom of Naples. The town of Isernia was destroyed, with a great portion of its inhabitants, in the terrible earthquake which so fearfully devastated the kingdom of Naples on the 26th of July, 1805, nineteen years after the appearance of the book alluded to. Perhaps with it perished the last trace of the worship of Priapus in this particular form, but Payne Knight was not acquainted with the fact that this superstition, in a variety of forms, prevailed throughout Southern and Western Europe largely during the Middle Ages, and that in some parts it is hardly extinct at the present day.

"The mediæval worship of the generative powers, represented by the generative organs, was derived from two distinct sources. In the first place, Rome invariably carried into the provinces she had conquered her own institutions and forms of worship, and established them permanently. In exploring the antiquities of these provinces, we are astonished at the abundant monuments of the worship of Priapus, in all the shapes and with all the attributes and accompaniments, with which we are already so well acquainted in Rome and Italy. Among the remains of Roman civilization in Gaul, we find statues or statuettes of Priapus, altars dedicated to him, the gardens and fields entrusted to his care, and the phallus, or male member, figured in a variety of shapes as a protecting power against evil influences of various kinds. With this idea the well-known figure was sculptured on the walls of public buildings, placed in conspicuous places in the interior of the house, worn as an ornament by women, and suspended as an amulet to the necks of children. Erotic scenes of the most extravagant description covered vessels of metal, earthenware, and glass, intended, no doubt, for festivals and usages more or less connected with the worship of the principle of fecundity."

At Aix, in Provence, there was found, on or near the site of the ancient baths, to which it had no doubt some relation, an enormous phallus, encircled with garlands, sculptured in white marble. At Le Chatelet, in Champagne, on the site of a Roman town, a colossal phallus was found. Similar objects in bronze, and of smaller dimensions, are so common, that explorations are seldom carried on upon a Roman site in which they are not found, and examples of such objects abound in the museums, public or private, of Roman antiquities.

At Nismes, in the south of France (formerly Nemansus), phallic worship appears to have prevailed to an unusual extent, and the walls of many of its buildings are adorned with sculptures of a most remarkable character, many of which can scarcely be regarded as other than fanciful caricatures. Some of them are thus described. One is the figure of a double phallus. It is sculptured on the lintel of one of the vomitories, or issues, of the second range of seats of the Roman amphitheatre, near the entrance gate which looks to the south. The double and the triple phallus are very common among the small Roman bronzes, which appear to have served as amulets, and for other similar purposes. In the latter, one phallus usually serves as the body, and is furnished with legs, generally those of the goat; a second occupies the usual place of this organ; and a third appears in that of a tail. On a pilaster of the amphitheatre of Nismes, we see a triple phallus of this description with goat's legs and feet. A bell is suspended to the smaller phallus in front; and the larger organ which forms the body is furnished with wings. The picture is completed by the introduction of three birds, two of which are pecking at the unveiled head of the principal phallus, while the third is holding down the tail-with its foot.

On the top of another pilaster of the amphitheatre at Nismes, to the right of the principal western entrance, was a bas-relief, also representing a triple phallus, with legs of a goat, and winged, but with a further accompaniment. A female, dressed in the Roman stola, stands upon the phallus forming the tail, and holds both it and the one forming the body with a bridle. This bas-relief was taken down in 1829, and is now preserved in the museum of Nismes.

A still more remarkable monument of this class was found in the course of excavations made at Nismes in 1825. It represents a bird, apparently intended for a vulture, with spread wings and phallic tail, sitting on four eggs, each of which is designed, no doubt, to represent the female organ. The local antiquaries give to this, as to the other similar objects, an emblematical signification; but it may be more rightly regarded, perhaps, as a playful conception of the imagination. A similar design, with some modifications, occurs not frequently among Gallo-Roman antiquities.

As Nismes was evidently a centre of this Priapic worship in the south of Gaul, so there appears to have been, perphaps lesser, centres in other parts, and we may trace it to the northern extremities of the Roman province, even to the other side of the Rhine. On the site of Roman settlements near Xanten, in Lower Hesse, a large quantity of pottery and other objects have been found, of a character to leave no doubt as to the prevalence of this worship in that quarter. But the Roman settlement which occupied the site of the modern city of Antwerp appears to have been one of the most remarkable seats of the worship of Priapus in the north of Gaul, and it continued to exist there till a comparatively modern period. *

So far as Britain is concerned, there is no doubt whatever that the worship of Priapus was established here as in other countries. Whenever there are any Roman remains of any particular extent, statuettes of Priapus, phallic bronzes, and pottery covered with obscene pictures are found. The bronzes found in England, are perfectly identical in character with those which occur in France and Italy. In illustration of this fact, we may mention two examples of the triple phallus, which appears to have been, perhaps in accordance with the explanation given by Plutarch, an amulet in great favour. One was found in London in 1842. As in examples found on the continent, a principal phallus formed the body, having the hinder parts of apparently a dog, with

* See Payne Knight's Work.

wings of a peculiar form, perhaps intended for those of a dragon. Several small rings are attached, probably for the purpose of suspending bells Another example was found at York in 1866, which displays a peculiarity of action, which leaves no doubt that the hinder parts were intended to be those of a dog.

Numerous examples of lamps of a phallic character have also been found in London and other parts connected with ancient Roman settlements. They are chiefly of earthenware, and some of them are of the most obscene description. One, found in London—in Cannon Street—in 1838, said to be typical of this class of objects, represents a natural act which will be readily understood.

One remarkable example of phallic monuments in Britain, demands, from its peculiarity, special notice. The author of the addition to Payne Knight's work says:—"All this obscene pottery must be regarded, no doubt, as a proof of a great amount of dissoluteness in the morals of Roman society in Britain, but it is evidence of something more. It is hardly likely that such objects could be in common use at the family table (this of course is in allusion to the lamps); and we are led to suppose that they were employed on special occasions, festivals perhaps, connected with the licentious worship of which we are speaking, and such as those described in the satires of Juvenal. But monuments are found in this island which bear still more direct evidence to the existence of the worship of Priapus during the Roman period.

"In the parish of Adel, in Yorkshire, are considerable traces of a Roman station which appears to have been a place of some importance, and which certainly possessed temples. On the site of these were found altars and other stones with inscriptions, which, after being long preserved in an outhouse of the rectory at Adel, are now deposited in the museum of the Philosophical Society at Leeds. One of the most curious of these appears to be a votive offering to Priapus, who seems to be addressed under the name of Mentula. It is a rough, unsquared stone, which has been selected for possessing a tolerably flat and smooth surface; and the figures and letters were made with a rude implement, and by an unskilful workman, who was evidently unable to cut a continuous smooth line. The middle of the stone is occupied by the figure of a phallus, and round it we read very distinctly the words:—**PRIMINVS MENTLA.** The author of the inscription may have been an ignorant Latinist as well as an unskilful sculptor, and perhaps mistook the ligulated letters, overlooking the limb which would make the L stand for VL, and giving A for Æ. It would then read *Priminus Mentula*, Priminus to Mentula (the object personified), and it may have been a votive offering from some individual named Priminus, who was in want of a heir or laboured under some sexual infirmity, to Priapus, whose assistance he sought. Another interpretation has been suggested, on the supposition that Mentla, or perhaps (the L being designed for IL ligulated) Mentila or Mentilla, might be the name of a female joined with her husband in this offering for their common good. The former of these interpretations seems, however, to be the most probable. This monument belongs probably to rather a late date in the Roman period. Another *ex voto* of the same class was found at Westerwood Fort, in Scotland, one of the Roman fortresses on the Wall of Antoninus. This monument consisted of a square slab of stone, in the middle of which was a phallus, and under it the words EX VOTO. Above were the letters XAN, meaning, perhaps, that the offerer had laboured ten years under the grievance of which he sought redress from Priapus. We may point also to a phallic monument of another kind, which reminds us in some degree of the finer sculptures at Nismes. At Housesteads, in Northumberland, are seen the extensive and imposing remains of one of the Roman stations on the Wall of Hadrian, named Borconicus. The walls of the entrance gateways are exceedingly well preserved, and on that of the guard-house attached to one of them is a slab of stone presenting a rude delineation of a phallus with the legs of a fowl, reminding us of some of the monuments of France and Italy. These phallic images were, no doubt, exposed in such situations because they were supposed to exercise a protective influence over the locality, or over the building, and the individual who looked upon the figure believed himself safe, during that day at least, from evil influences of various descriptions."

The Chronicle of Lanercost supplies us with some curious information relative to Phallic practices in Britain some five or six hundred years ago. A pestilence we are told prevailed in the Scottish district of Lothian, which was very fatal to the cattle, and to counteract which some of the clergy taught the peasantry to make a fire by the rubbing together of wood (this was the need-fire), and to raise up the image of Priapus as a means of saving their cattle. When a lay member of the Cistercian order, at Fenton, had done this before the door of the hall, and had sprinkled the cattle with a dog's testicles dipped in holy water, and complaint had been made of this crime of idolatry against the lord of the manor, the latter pleaded in his defence that all this was done without his knowledge and in his absence, but added, "while until the present month of June, other people's cattle fell ill and died, mine were always sound, but now every day two or three of mine die, so that I have few left for the labours of the field."

In 1282, a similar case occurred at Inverkeithing, in the county of Fife, Scotland. In the Easter week of this year, a parish priest, named John, performed the rites of Priapus, by collecting the young girls of the town, and making them dance round the figure of this god; without any regard for the sex of these worshippers, he carried a wooden image of the male members of generation before them in the dance, and himself dancing with them, he accompanied their songs with monuments in accordance, and urged them to licentious actions by his licentious language. The more modest part of those who were present felt scandalized by the proceedings, and expostulated with the priest, but he treated their words with contempt, and only gave utterance to coarser obscenities. He was cited before his bishop, defended himself upon the common usage of the country, and was allowed to retain his benefice.

In the Middle Ages, the buildings which were specially placed under the supposed protective power of the phallus sculptured on the walls, were generally churches. In Ireland, singularly enough, it was the female organ which was regarded as the protection against enchantments of various kinds, and which to an ignorant and superstitious people were a source of so much terror. These images were, as pictures of them show, of an extremely rude, though elaborate description, and indecent in the highest degree; they were placed generally over the doorway of the church, and represented women exposing themselves in the most outrageous manner by uncovering those parts which, as a rule, are concealed even by the most savage of people. It seems tolerably clear that we have here the origin of the custom, which has descended to our times, of nailing up a horse-shoe as a protection against the evil influence of witches. It has been observed that this female organ was "far more liable to degradation in form than that of the male, because it was much less easy, in the hands of rude draughtsmen, to delineate an intelligible form, and hence it soon assumed shapes which, though intended to represent it, we might rather call symbolical of it, though no symbolism was intended. Thus the figure of the female organ easily assumed the rude form of a horse-shoe, and, as the original meaning was forgotten, would be readily taken for that object, and a real horse-shoe nailed up for the same purpose."

A curious aspect of this matter is presented in the middle ages by the conversion of the god Priapus into a saint of several names. In the south of France, Provence, Languedoc, and some other places, he was worshipped as St. Foutin, a name said to be a corruption of Fotinus, the first bishop of Lyons. The object of adoration was a large phallus of wood, this was greatly venerated by the women, who scraped it with knives and swallowed the fragments in water as a cure for barrenness on their own part, and gave them to their husbands as a remedy for any incapacity or weakness in fulfilling the obligations of the married state. Volume five of the Journal d'Henri III., containing "La Confession de Sancy," gives an account of the worship of this saint in France at the commencement of the seventeenth century. That work states that at Varailles, in Provence, waxen images of the members of both sexes were offered to St. Foutin, and suspended from the ceiling of his chapel. At Ermbrun, in the department of the Upper Alps, the phallus of St. Foutin was worshipped in a different form. The women poured a libation of wine upon its head, which was collected in a vessel, in which it was left till it became sour; it was then called the "sainte vinaigre," and was used for a purpose not quite clearly comprehended. In 1585, when Erubrum was taken by the Protestants, this phallus was found carefully laid up among the relics of the principal church, its head red with the wine that had been poured upon it. At Orange, in the church of St. Eutropius, a very large phallus of wood, covered with leather, was taken possession of by the Protestants in 1562, and burnt. The same name (St. Foutin) is found in connection with worship in a number of other places, such as Porigny, Cives, Vendre in the Bourbonnais, Auxerre, Puy-en-Velay, &c. A number of other phallic saints were also worshipped in the middle ages, such as St. Guerlichon at Bourg-Dieu, St. Gilles in the Cotentin in Brittany, St. René in Anjou, St. Regnaud in Burgundy, and St. Guignolé, near Brest, and at the village of La Chatelette in Berri.

"It appears that it was also the practice to worship these saints in another manner, which was also derived from the forms of the worship of Priapus among the ancients, with whom it was the custom, in the nuptial ceremonies, for the bride to offer up her virginity to Priapus, and this was done by placing her sexual parts against the end of the phallus, and sometimes introducing the latter, and even completing the sacrifice. This ceremony is represented in a bas-relief in marble, an engraving of which is given in the Musée Sécret of the antiquities of Herculaneum and Pompeii; its object was to conciliate the favour of the god, and to avert sterility. It is described by the early Christian writers, such as Lactantius and Arnobius, as a very common practice among the Romans, and it still prevails to a great extent over most part of the East, from India to Japan, and the islands of the Pacific. In a public square in Batavia, there is a cannon taken from the natives, and placed there as a trophy by the Dutch Government. It presents the peculiarity that the touch-hole is made on a phallic hand, the thumb placed in the position which is called the "fig." At night, the sterile Malay women go to this cannon and sit upon the thumb, and rub their parts with it to produce fruitfulness. When leaving, they make an offering of a bouquet of flowers to the "fig." It is always the same idea of reverence to the fertilizing powers of nature, of which the garland or the bunch of flowers was an appropriate emblem. There are traces of the existence of this practice in the middle ages. In the case of some of the priapic saints mentioned above, women sought a remedy for barrenness by kissing the end of the phallus; sometimes they appear to have placed a part of their body naked against the image of the saint, or to have sat upon it. The latter trait was perhaps too bold an adoption of the indecencies of pagan worship to last long, or to be practised openly; but it appears to have been more innocently represented by lying upon the body of the saint, or sitting upon a stone, understood to represent him without the presence of the energetic member. In a corner in the church of the village of St. Fiacre, near Monceaux, in France, there is a chair called the chair of St. Fiacre, which confers (so it is said) fecundity upon women who sit upon it, but it is necessary that nothing should intervene between their bare skin and the stone." *

* *Worship of Priapus.*

Antwerp has been described as the Lampsacus of Belgium, and Priapus was, down to a comparatively modern period, its patron saint, under the name of Ters, a word, the derivation of which appears to be unknown, but which was identical in meaning with the Greek *phallus*, and the Latin *fascinum*. Johannis Goropii Becani *Origines Antwerpianæ*, 1569, lib. I., pp. 26, 101, informs us how much this Ters was reverenced in his time by the Antwerpians, especially by the women, who invoked it on every occasion when they were taken by surprise or sudden fear. He states that if they let fall by accident a vessel of earthenware, or stumbled, or if any unexpected accident caused them vexation, even the most respectable woman called aloud for protection of Priapus under this obscene name. Goropius Becanus adds that there was in his time, over the door of a house adjoining the prison, a statue which had been furnished with a large phallus, then worn away or broken off. Among other writers who mention this statue is Abraham Golnitz, who published an account of his travels in France and Belgium, in 1631, and he informs us that it was a carving in stone, about a foot high, with its arms raised up, and its legs spread out, and that the phallus had been entirely worn out by the women, who had been in the habit of scraping it and making a potion of the dust, which they drank as a preservative against barrenness. Golnitz further tells us that a figure of Priapus was placed over the entrance-gate to the enclosure of the temple of St. Walburgis, at Antwerp, which some antiquaries imagined to have been built on the site of a temple dedicated to that deity. It appears from these writers that, at certain times, the women of Antwerp decorated the phalli of these figures with garlands.

We have seen, says the author of "The Worship of the Generative Powers," how the women of Antwerp, who, though perhaps they did not speak a Roman dialect, appear to have been much influenced by Roman sentiments, made their appeal to their genius Ters. When a Spaniard is irritated, or suddenly excited, he exclaims, *Carajo!* (the virile member), or *Conjones!* (the testicles). An Italian, under similar circumstances, uses the exclamation *Cazzo!* (the virile member). The Frenchman apostrophizes the act, *Foutre!* The female member, *cono* with the Spaniard, *conno* with the Italian, and *con* with the Frenchmen, was and is used more generally as an expression of contempt, which is also the case with the testicles, *couillons*, in French—

those who have had experience in the old days of "diligence" travelling will remember how usual it was for the driver, when his horses would not go quick enough, to address the leader in such terms as, "Va, donc, vieux con!" We have no such words used in this manner in the Germanic languages, with the exception, perhaps, of the German. *Potz!* and *Potztavsend!* and the English equivalent, *Pox!* which last is quite out of use. There was an attempt among the fashionables of our Elizabethan age of literature, to introduce the Italian *cazzo* under the form of *catso*, and French *foutre* under that of *foutra*, but these were mere affectations of a moment, and soon disappeared.

CHAPTER VII.
Moral Aspects of Phallicism.

OF all the profane rites which belonged to the ancient polytheism, none were more furiously inveighed against by the zealous propagators of the Christian faith, than the obscene ceremonies performed in the worship of Priapus, which appeared not only contrary to the gravity and sanctity of religion, but subversive of the first principles of decency and good order in society. Even the form itself, under which the god was represented, appeared to them a mockery of all piety and devotion, and more fit to be placed in a brothel than a temple. But the forms and ceremonials of a religion are not always to be understood in their direct and obvious sense; but are to be considered as symbolical representations of some hidden meaning, which may be extremely wise and just, though the symbols themselves, to those who know not their true signification, may appear in the highest degree absurd and extravagant. It has often happened, that avarice and superstition have continued these symbolical representations for ages after their original meaning has been lost and forgotten; when they must of course appear nonsensical and ridiculous, if not impious and extravagant.

Such is the case with the rite now under consideration, than which nothing can be more monstrous and indecent, if considered in its plain and obvious meaning, or as a part of the Christian worship, but which will be found to be a very natural symbol of a very natural and philosophical system of religion, if considered according to its original use and intention.

Whatever the Greeks and Egyptians meant by the symbol in question, it was certainly nothing ludicrous or licentious; of which we need no other proof than its having been carried in solemn procession at the celebration of those mysteries in which the first principles of their religion, the knowledge of the God of Nature, the First, the Supreme, the Intellectual, were preserved free from the vulgar superstitions, and communicated, under the strictest oaths of secrecy, to the initiated, who were obliged to purify themselves, prior to their initiation, by abstaining from venery and all impure food. We may therefore be assured that no impure meaning could be conveyed by this symbol, but that it represented some fundamental principle of their faith. What this was, it is difficult to obtain any direct information, on account of the secrecy under which this part of their religion was guarded. Plutarch tells us that the Egyptians represented Osiris with the organ of generation erect, to show his generative and prolific power; he also tells us that Osiris was the same deity as the Bacchus of the Greek mythology, who was also the same as the first-begotten Love of Orpheus and Hesiod. This deity is celebrated by the ancient poets as the creator of all things, the father of gods and men, and it appears by the passage above referred to that the organ of generation was the symbol of his great characteristic attribute. This is perfectly consistent with the general practice of the Greek artists, who uniformly represented the attributes of the deity by the corresponding properties observed in the objects of sight. They thus personified the epithets and titles applied to him in the hymns and litanies, and conveyed their ideas of him by forms, only intelligible to the initiated, instead of sounds which were intelligible to all. The organ of generation represented the generative or creative attribute, and in the language of painting and sculpture, signified the same as the epithet $\pi\alpha\gamma\gamma\epsilon\nu\epsilon\tau\omega\rho$, in the Orphic litanies.

This interpretation will perhaps surprise those who have not been accustomed to divest their minds of the prejudices of education and fashion; but I doubt not but it will appear just and reasonable to those who consider manners and customs as relative to the natural causes which produced them rather than to the artificial opinions and prejudices of any particular age or country. There is naturally no impurity or licentiousness in the moderate and regular gratification of any natural appetite; the turpitude consisting wholly in the excess or perversion. Neither are organs of one species of enjoyment to be considered as subjects of shame and concealment more than those of another, every refinement on this head being derived from acquired habit, not from nature; habit indeed long established, for it seems to have been as general in Homer's days as at present, but which certainly did not exist when the mystic symbols of the ancient worship were first adopted. As these symbols were intended to express abstract ideas by objects of sight, the contrivers of them naturally selected those objects whose characteristic properties seemed to have the greatest analogy with the divine attributes which they wished to represent.

In an age, therefore, when no prejudices of artificial decency existed, what more just and natural image could they find by which to express their idea of the beneficial power of the great Creator than that organ which endowed them with the power of procreation, and made them partakers not only of the felicity of the Deity, but of his great characteristic attribute, that of multiplying his own image, communicating his blessings, and extending them to generations yet unborn.

Mr. d'Hancarville attributes the origin of all these symbols to the antiquity of words; the same term being employed in the primitive language to signify God and a Bull, the Universe and a Goat, Life and a Serpent. But words are only types and symbols of ideas, and therefore must be posterior to them, in the same manner as ideas are to the objects. The words of a primitive language, being imitative of the ideas from which they spring, and of the objects they meant to express, as far as the imperfections of the organs of speech will admit, there must necessarily be the same kind of analogy between them as between the ideas and objects themselves. It is impossible, therefore, that in such a language any ambiguity of this sort could exist, as it does in secondary language; the words of which, being collected from various sources, and blended together without having any natural connection, become arbitrary signs of convention, instead of imitative representatives of ideas. In this case it often happens, that words, similar in form, but different in meaning, have been adopted from different sources, which, being blended together, lose their little difference of form, and retain their entire difference of meaning. Hence ambiguities arise, such as those above mentioned, which could not possibly exist in an original language.

Considering the general state of reserve and restraint in which the Grecian women lived, it is astonishing to what an excess of extravagance their religious enthusiasm was carried on certain occasions; particularly in celebrating the Orgies of Bacchus. The gravest matrons and proudest princesses suddenly laid aside their decency and dignity, and ran screaming among the woods and mountains, fantastically dressed or half-naked, with their hair dishevelled and interwoven with ivy or vine, and sometimes with living serpents. In this manner they frequently worked themselves up to such a pitch of savage ferocity, as not only to feed upon raw flesh, but even to tear living animals with their teeth, and eat them warm and palpitating. Mr. Payne Knight says:—" The intelligent reader perceives the superficiality of the popular notion that Bacchus or Dionysus was but the god of wine and drunkenness, and that the orgies or secret religious rites were all occasions of revelling and debauchery. His worshippers in Thrace, the Orpheans, were ascetics and devotees, like the Gymnosophists of India. The Bacchus of ancient worship was an Asiatic divinity, identical with Atys, Adonis, Osiris, and probably with Maha Deva of India; and in the Grecian Pantheon he appears to be a foreigner like Hercules. As Zagreus, the son of Zeus by the Virgin Kore-Persephoneia or Demeter, afterwards born anew as the son of Semelê, he seems to illustrate the metempsychosis. He was probably identical with Baal-Peor, the Moabite divinity, and the deity commemorated by the Israelites in the 'Baalim' or Priapic statues, often of wood, which were set up with the groves or symbols of Venus-Astartê, on every high hill and under every green tree. Maachah, the queen-mother, who presided over the orgies, was deposed from regal rank by King Asa for making a mephallitzeth, or phallic mannikin for an ashera, or omphale. (I. Kings, xv. 13, and Herodotus, ii. 48). The orgies, works, or nocturnal rites, consisted of dances, mystical processions, and searches after the mutilated body of the divine youth."

One great means of corrupting the ancient theology, and establishing the poetical mythology, was the practice of the artists in representing the various attributes of the Creator under human forms of various character and expression. These figures, being distinguished by the titles of the deity which they were meant to represent, became in time to be considered as distinct personages, and worshipped as separate subordinate deities. Hence the many-shaped god, the πολυμορφος and μυριομοεος of the ancient theologists, became divided into many gods and goddesses, often described by the poets as at variance with each other, and wrangling about the little intrigues and passions of men. Hence too, as the symbols were multiplied, particular ones lost their dignity; and that venerable one which is the subject of our consideration, became degraded from the representative of the god of nature to a subordinate rural deity, a supposed son of the Asiatic conqueror Bacchus, standing among the nymphs by a fountain, and expressing the fertility of a garden, instead of the general creative power of the great active principle of the universe. His degradation did not stop even here; for we find him, in times still more profane and corrupt, made a subject of raillery and insult, as answering no better purpose than holding up his rubicund nose to frighten birds and thieves.* His talents were also perverted from their natural ends, and employed in base and abortive efforts in conformity to the taste of the times; for men naturally attribute their own passions and inclinations to the objects of their adoration.

"The Babylonians," says Herodotus (Book I. c 199), "have a most shameful custom. Every woman born in the country must once in her life go and sit down in the precinct of Venus, and there consort with a stranger. Many of the wealthier sort, who are too proud to mix with the others, drive in covered carriages to the precinct, followed by a goodly train of attendants, and there take their station. But the larger number seat themselves within the holy enclosure with wreaths of string about their heads,—and here there is always a great crowd, some coming and others going; lines of cord mark out paths in all directions among the women, and the strangers pass along them to make their choice. A woman who has once taken her seat is not allowed to return home till one of the strangers throws a silver coin into her lap, and takes her with him beyond the holy ground. When he throws the coin he says these words—'the goddess Mylitta prosper thee.' (Venus is called Mylitta by the Assyrians). The silver coin may be of any size; it cannot be refused, for that is forbidden by the law, since once thrown it is sacred. The woman goes with the first man who throws her money, and rejects no one. When she has gone with him, and so satisfied the goddess, she returns home, and from that time forth no gift, however great, will prevail with her. Such of the women, as are tall and beautiful are soon released, but others who are ugly have to stay a long time before they can fulfil the law. Some have waited three or four years in the precinct."

There is an allusion to this in the Book of Baruch (vi. 43). "The women also with cords about them, sitting in the ways, burn bran for perfume: but if any of them, drawn by some that passeth by, lie with him, she reproacheth her fellow, that she was not thought as worthy as herself, nor her cord broken."

A similiar custom prevailed in Cyprus, Armenia, and probably in many other countries; it being, as Herodotus observes, the

* Horat. lib. i. Sat. VIII.

practice of all mankind, except the Greeks and Egyptians, to take such liberties with their temples, which, they concluded, must be pleasing to the Deity, as birds and animals, acting under the guidance of instinct, or by the immediate impulse of Heaven, did the same.

Thus Herodotus ii. 64:—"The Egyptians first made it a point of religion to have no converse with women in the sacred places, and not to enter them without washing, after such converse. Almost all other nations, except the Greeks and Egyptians, act differently, regarding man as in this matter under no other law than the brutes. Many animals, they say, and various kinds of birds may be seen to couple in the temples and the sacred precincts, which would certainly not happen if the gods were displeased at it."

Payne Knight says:—"The exceptions he might safely have omitted, at least as far as relates to the Greeks: for there were a thousand sacred prostitutes kept in each of the celebrated temples of Venus, at Eryx and Corinth; who, according to all accounts, were extremely expert and assiduous in attending to the duties of their profession; and it is not likely that the temple, which they served, should be the only place exempted from being the scene of them. Dionysius of Halicarnasus claims the same exception in favour of the Romans, but, as we suspect, equally without reason: for Juvenal, who lived only a century later, when the same religion and nearly the same manners prevailed, seems to consider every temple in Rome as a kind of licensed brothel."

The temples of the Hindus, in the Dekkan, possessed their establishments; they had bands of consecrated dancing-girls called the Women of the Idol, selected in their infancy by the priests for the beauty of their persons, and trained up with every elegant accomplishment that could render them attractive, and assure success in the profession; which they exercised at once for the pleasure and profit of the priesthood. They were never allowed to desert the temple; and the offspring of their promiscuous embraces were, if males, consecrated to the service of the Deity in the ceremonies of his worship; and if females, educated in the profession of their mothers.

"The compulsory prostitution of Babylonia was connected with the worship of Mylitta, and wherever this worship spread it was accompanied by the sexual sacrifice. Strabo relates that in

Armenia the sons and daughters of the leading families were consecrated to the service of Anaitis for a longer or shorter period. Their duty was to entertain strangers, and those females who had received the greatest number, were, on their return home, the most sought after in marriage. The Phœnician worship of Astarté was no less distinguished by sacred prostitution, to which was added a promiscuous intercourse between the sexes during certain religious fêtes, at which the men and women exchanged their garments. The Phœnicians carried the customs to the Isle of Cyprus, where the worship of their great goddess, under the name of Venus became supreme.

"According to a popular legend the women of Amathonte, afterwards noted for its temple, were originally known for their chastity. When, therefore, Venus was cast by the waves naked on their shores, they treated her with disdain, and, as a punishment, they were commanded to prostitute themselves to all comers, a command which they obeyed with so much reluctance that the goddess changed them into stone. With their worship of Astarté, or Venus, the Phœnicians introduced sacred prostitution into all their colonies. St. Augustine says that, at Carthage, there were three Venuses rather than one: one of the virgins, another of married women, and a third of the courtesans, to the last of whom it was that the Phœnicians sacrificed the chastity of their daughters before they were married. It was the same in Syria. At Byblos, during the fêtes of Adonis, after the ceremony which announced the resurrection of the god, every female worshipper had to sacrifice to Venus either her hair or her person. Those who preferred to preserve the former adjourned to the sacred enclosures, where they remained for a whole day for the purpose of prostituting themselves."

A similar state of things prevailed in many other countries, and, according to Herodotus, particularly amongst the Lydians. He says in his first book, Lydia has one structure of enormous size, only inferior to the monuments of Egypt and Babylon.* This is the tomb of Alyattes, the father of Crœsus, the base of which is formed of immense blocks of stone, the rest being a vast mound of earth. It was raised, he says, by the joint labour of the tradesmen, handicraftsmen, and courtesans of Sardis, and had at the top fine stone pillars, which remained to his day, with inscriptions cut on them, shewing how much of the work was

* C. S. Wake's *Essays*.

done by each class of workpeople. It appeared on measurement that the portion of the courtesans was the largest. He then remarks that the daughters of the common people in Lydia, one and all, pursued this traffic in order to collect money for their portions, and that they continued the practice till they married.

A note in Rawlinson's translation says the statement about the Lydian women is one of those for which Herodotus cannot escape censure, and other writers have denied the practice of sacred prostitution by the Egyptians; "the great similarity, however, between the worship of Osiris and Isis and that of Venus and Adonis," remarks Wake, "renders the contrary opinion highly probable." Dufour declares that the females on their way to the fêtes of Isis at Bubastis, executed indecent dances when the vessels passed the villages on the banks of the river, and that these obscenities were only such as were about to happen at the temple, which was visited each year by seven hundred thousand pilgrims, who gave themslves up to incredible excesses, and Strabo affirms that a class of persons called pellices (harlots) were dedicated to the service of the patron saint of Thebes, and that they were permitted to cohabit with anyone they chose.

Numerous references to similar practices are found in other writers, bearing more or less directly upon our subject. Xenophon consecrated fifty courtesans to the Corinthian Venus, in pursuance of the vow which he had made when he besought the goddess to give him the victory in the Olympian games, and he thus addresses them:—"Oh, young damsels, who receive all strangers and give them hospitality, priestesses of the goddess Pitho in the rich Corinth, it is you who, in causing the incense to burn before the images of Venus, and in invoking the mother of Love, often merit for us her celestial aid, and procure for us the sweet moments which we taste on the luxurious couches where is gathered the delicate fruit of beauty."

Clement of Alexandria, in his "Exhortation to the Heathen," has some very striking allusions to the phallic mysteries as practised in his time, chiefly in the countries of Egypt and Greece. He is writing respecting the absurdity of those mysteries, and amongst his many denunciations says:—"There is then the foam-born and Cyprus-born, the darling of Cinyras,—I mean Aphrodite, lover of the virilia, because sprung from them, even from those of Uranus, that were cut off,—those lustful members that, after being cut off, offered violence to the waves. Of members so lewd a worthy fruit—Aphrodite—is born. In the rites which celebrate this enjoyment of the sea, as a symbol of her birth a lump of salt and the phallus are handed to those who are initiated into the art of uncleanness. And those initiated bring a piece of money, as a courtesan's paramours do to her.

"Then there are the mysteries of Demeter, and Zeus's wanton embraces of his mother, and the wrath of Demeter; I know not what for the future I shall call her, wife or mother, on which account it is that she is called Brimo, as is said; also the entreaties of Zeus, and the drink of gall, the plucking out of the hearts of sacrifices, and deeds that we dare not name. Such rites the Phrygians perform in honour of Attis and Cybele and the Corybantes. And the story goes, that Zeus, having torn away the testicles of a ram, brought them out and cast them at the breasts of Demeter, paying thus a fraudalent penalty for his violent embrace, pretending to have cut out his own. The symbols of initiation into these rites, when set before you in a vacant hour, I know will excite your laughter, although on account of the exposure by no means inclined to laugh. I have eaten out of the drum, I have drunk out of the cymbal, I have carried the Cernos, I have slipped into the bedroom. Are not these signs a disgrace? Are not the mysteries absurdity?

"What if I add the rest? Demeter becomes a mother, Kore is reared up to womanhood. And, in course of time, he who begot her—this same Zeus—has intercourse with his own daughter Pherephatta,—after Ceres, the mother—forgetting his former abominable wickedness. Zeus is both the father and the seducer of Kore, and has intercourse with her in the shape of a dragon; his identity, however, was discovered. The token of the Sabazian mysteries to the initiated is "the deity gliding over the breast"— the deity being this serpent crawling over the breasts of the initiated. Proof surely this of the unbridled lust of Zeus. Pherephatta has a child, though, to be sure, in the form of a bull, as an idolatrous poet says:

> 'The bull
> The dragon's father, and father of the bull the dragon,
> On a hill the herdsman's hidden ox-goad,'—

alluding, as I believe, under the name of the herdsman's ox-goad, to the reed wielded by the bacchanals. Do you wish me to go into the story of Pherephatta's gathering of flowers, her basket, and her seizure by Pluto (Aidoneus), and the rent in the earth, and the swine of Eubouleus that were swallowed up with the two goddesses; for which reason, in the Thesmophoria, speaking the Megaric tongue, they thrust out swine! This mythological story the women celebrate variously in different cities in the festivals called Thesmophoria and Scirophoria, dramatizing in many forms the rape of Pherephatta (Proserpine).

"The mysteries of Dionysus are wholly inhuman: for while still a child, and the Curetes danced around [his cradle] clashing their weapons, and the Titans having come upon them by stealth and having beguiled him with childish toys, these very Titans tore him limb from limb when but a child, as the bard of this mystery, the Thracian Orpheus, says:

"Cone, and spinning-top, and limb-moving rattles,
And fair golden apples from the clear-toned Hesperides."

"And the useless symbols of this mystic rite it will not be useless to exhibit for condemnation. These are dice, ball, hoop, apples, top, looking-glas, tuft of wool.

"Athene (Minerva), to resume our account, having abstracted the heart of Dionysus, was called Pallas, from the vibrating of the heart, and the Titans who had torn him limb from limb, setting a cauldron on a tripod, and throwing into it the members of Dionysus, first boiled them, and then, fixing them on spits, held them over the fire. But Zeus having appeared, since he was a god, having speedily perceived the savour of the pieces of flesh that were being cooked,—that savour which your gods agree to have assigned to them as their perquisite—assails the Titans with his thunderbolt, and consigns the members of Dionysus to his son Apollo to be interred. And he—for he did not disobey Zeus—bore the dismembered corpse to Parnassus, and there deposited it.

"If you wish to inspect the orgies of the Corybantes, then know that, having killed their third brother, they covered the head of the dead body with a purple cloth, crowned it, and carrying it on the point of a spear, buried it under the roots of Olympus. These mysteries are, in short, murders and funerals. And the priests of these rites, who are called kings of the sacred rites by those whose business it is to name them, give additional strangeness to the tragic occurrence, by forbidding parsley with the roots from being placed on the table, for they think that parsley grew from the Corybantic blood that flowed forth; just as the women, in celebrating the Thesmophoria, abstain from eating the seeds of the pomegranate which have fallen on the ground, from the idea that pomegranates sprang from the drops of the blood of Dionysus. Those Corybantes also they call Cabiri; and the ceremony itself they announce as the Cabiric mystery.

"For those two identical fratricides, having abstracted the box in which the member of Bacchus was deposited, took it to Etruria—dealers in honourable wares truly. They lived there as exiles, employing themselves in communicating the precious teaching of their superstition, and presenting the genitals and the box for the Tyrrhenians to worship. And some will have it, not improbably, that for this reason Dionysus was called Attis, because he was castrated. And what is surprising, at the Tyrrhenians, who were barbarous, being thus initiated into these foul indignities, when among the Athenians, and in the whole of Greece—I blush to say it—the shameful legend about Demeter holds its ground? For Demeter, wandering in quest of her daughter Core, broke down with fatigue near Eleusis, a place in Attica, and sat down on a well, overwhelmed with grief. This is even now prohibited to those who are initiated, lest they should appear to mimic the weeping goddess. The indigenous inhabitants then occupied Eleusis: their names were Baubo, and Dusaules, and Triptolemus; and besides, Eumolpus and Eubouleus. Triptolemus was a herdsman, Eumolpus a shepherd, and Eubouleus a swineherd; from whom came the race of the Eumolpidæ and that of the Heralds—a race of Hierophants—who flourished at Athens.

"Well, then (for I shall not refrain from the recital), Baubo having received Demeter hospitably, reaches to her a refreshing draught, and on her refusing it, not having any inclination to drink (for she was very sad), and Baubo having become annoyed, thinking herself slighted, uncovered, and exhibited herself to the goddess. Demeter is delighted at the sight, and takes, though with difficulty, the draught—pleased, I repeat, at the spectacle. These are the secret mysteries of the Athenians; these Orpheus records. I shall produce the very words of Orpheus, that you may have the great authority on the mysteries himself, as evidence for this piece of turpitude:

"Having thus spoken, she drew aside her garments,
And showed all that shape of the body which it is improper to name, the growth of puberty;
And with her own hand Baubo stripped herself under the breasts.
Blandly then the goddess laughed and laughed in her mind,
And received the glancing cup in which was the draught."

And the following is the token of the Eleusinian mysteries: *I have fasted, I have drunk the cup; I have received from the box; having done, I put it into the basket, and out of the basket into the chest.* Fine sights truly, and becoming a goddess; mysteries worthy of the night, and flame, and the magnanimous or rather silly people of the Erechthidæ, and the other Greeks besides, 'whom a fate they hope not for awaits after death.' And in truth against these Heraclitus the Egyptian prophesies, as 'the nightwalkers, the magi, the bacchanals, the Lenæan revellers, the initiated.' These he threatens with what will follow death, and predicts for them fire. For what are regarded among men as mysteries, they celebrate sacrilegiously. Law, then, and opinion, are nugatory. And the mysteries of the dragon are an imposture, which celebrates, religiously, mysteries that are no mysteries at all; and observes with a spurious piety profane rites. What are these mystic chests?—for I must expose their sacred things, and divulge things not fit for speech. Are they not sesame cakes, and pyramidal cakes, and globular and flat cakes, embossed all over, and lumps of salt, and a serpent the symbol of Dionysus Bassareus? And besides these, are there not pomegranates, and branches, and rods, and ivy leaves? and besides round cakes and poppy seeds? And further, there are the unmentionable symbols of Themis, marjoram, a lamp, a sword, a woman's comb, which is a euphemism and mystic expression for a woman's secret parts."

CHAPTER VIII.

Sacti Puga, the Worship of the Female Power.

THE two influences, Male and Female, are conspicuous in certain differences in the Phallic monuments, which unitedly, however, signify the same thing. The disputes of the comparative superiority of the Male over the Female principle, or of the Female over the Male, were the origin amongst the earliest nations of vast desolatory wars of which no history, scarcely even legend, has descended to modern periods. Therefore no accounts remain of these primeval wars, which brought about the building of the famous Tower of Babel, and were ultimately the cause of the confusion of languages, and the original dispersion of the nations. Obelisks, towers, and steeples represent and figure forth the Male principle. Pyramids, circular magnified forms, and rhomboidal or undulating serpentine shapes, denote the Female natural power. The one set of forms are masculine: therefore, aggressive and compelling. The other set of forms are feminine: therefore, submissive and ennobling. But all are alike phallic, and mean the same thing, that is, the natural motived power which causes and directs the world, that power which *is* the world, in fact.

Phallic objects, innumerable, are always peculiar in their form, and are of all sizes. If these sometimes prodigious structures are Obelisks, Columns, or Pillars, or as occasionally happens, simple, rough hewn, or partly fashioned uprights, they represent and figure forth the male principle. Subsequent to the very early, devotional ages, these pins, or uprights, assumed the forms of solid or slender towers, tors, or springing, rising, pointed fabrics. Amongst the Muslemmins these were minarets, with egg-shaped summits; in the architectural practice among the Christians, the tower attenuated into the spire or steeple. But the memorial structures with the larger base, and with that broader incidence which might be denominated, with a certain aptness, the Saturnian angle, indicates the opposite influence, that of the female, in mystic type or apotheosis. These symbol-structures, involving the idea of the feminine power, are the more broadly vaulting in shape. Chief, and most majestic of all these monuments, are the

NATURE WORSHIP.

Pyramids. All the mystic monuments of this form and fashion are in the general sense, equally Phalli—that is, devoted to, and in witness of the worship of the distinctive sexual peculiarities. We accept the whole as meaning the one thing, Phallicism, all interpreted under the general, rising, forceful form, aspiring towards the stars. Stately beyond idea, and gloomily majestic, as is the aspect of these lunar or womanly monumental structures, they can be soon distinguished. This group of the Feminine-Phallic forms comprises the Pyramids in the first rank. The Obleisk is a shrunken, vertically thin, concentrated pyramid : the Pyramid is a widely squared out obelisk, both express the same idea. In the conveyance of certain ideas to those who comtemplate them, the pyramid boasts of prouder significance, and impresses with a hint of still more impenetrable and more removed mystery. We seem to gather dim, supernatural ideas of the mighty mother of nature, the dusk divinity crowned with towers, the ancientest among the ancients, the Isis, or mysterious consort of the dethroned, and ruined, that almost two-sexed entity without a name. She of the Veil which is never to be lifted, perhaps not even by the angels, for their knowledge is limited. In short, this tremendous abstraction, Cybele, *Ideæ Mater*, Isiac controller of the Zodiacs, whatever she be, has her representative figure in the half-buried Sphynx even to our own day, watching the stars, although nearly swallowed up in the engulfing sands. This is the Gorgon survival of the period of the Ark, eldest daughter of the mythologies, whose other face (for Janus-like she looks two ways), turned away from the world, is beautiful as the fairest one of Paradise. That other face of the Gorgon, or Sphynx, resembles, in one respect, that side of the lunar disc ; the side of the moon turned away from observers on the earth, that face which no mortal eye ever saw, or will see, and which, for this reason, is one of the greatest mysteries in all the sky.

The foregoing remarks furnish the clue to this double history of the phallus, in the divided character of its worship, whether the obelisk or pillar, or whether the pyramid be the idol. It is too plain to be misunderstood. As the Greeks wrote *Palai* for Páli, they rendered the word Paliputra by *Palaigonos*, which also means the offspring of Pali, literally signifying the offspring of the phallus. It was notoriously the yoni, and not the phallus, which alone received the veneration of the Hindus, though now divided into innumerable sects and an inextricable mass of polytheism. Wilford observes that the Yávanas were the ancestors of the Greeks, and that the Pandits insist that the words Yávana and Yóni are derived from the root, Yu, and that the Yávanas were so named from their obstinate assertion of a superior influence in the female over the male nature. An ancient book on astronomy, in Sanscrit, bears the title of Yávana Jatici, which may be interpreted, "the Ionic sect." There is an ancient proverb amongst the Pandits, that "no base creature can be lower than a Yavana," truly showing the fluctuating nature of human opinions and theories, which, nevertheless, have torn the bosom of society, and shaken nations to their centre. This creed caused the new people in Greece to name their new country itself Ionia, from that consecrated Yoni which they revered, and to distinguish themselves as the Ionic, or Yónic sect, in indubitable reference to their peculiar opinions. These and such-like researches furnish us with the real meaning of proper names, and amongst others that of the great goddess Juno, which Wilford asserts to be derived from the Yoni of the Hindus; also, if we analyse the name of Diana, or Di-Yana, the great goddess of the Ephesians, we shall at once perceive an identical etymology; and when we remember that Juno was fabled to have been born at Argos, and that she was peculiarly worshipped there, we shall fully coincide in that opinion, for it is to be observed that the name of Argha is derived from the Bhaga of the Hindus, and both signified the Yoni, and likewise an ark or boat, which was used throughout antiquity as a type of the Yoni itself. The Hindu goddess, Bagis, was indifferently called Vagis, from which, no doubt, is derived the Latin *vagina;* and when we remember that Plutarch makes the otherwise inexplicable assertion that Osiris (or the incarnation of the male principle) was commander of the Argo; and when we learn that the true meaning of the name Argha-nátha, or, as we mostly render it (speaking of the great idol), Jagernath, is no other than "lord of the boat," we shall perceive at once the drift of these dark sentences, when truly and intelligibly expounded.

The discussion of this word Argha naturally induces us to remark concerning an intermediate or middle sect, which, says Wilford, "is now prevalent in India, and which was generally diffused over ancient Europe." It was introduced by the Pelargi, who, Herodotus says, were the same as the Pelasgi. Many ancient writers affirm that they were one of the most ancient peoples in the world. It is asserted that they first inhabited

Argolis, and about 1883, B.C,. passed into Æmonia, or Yomonia, and were afterwards dispersed, or emigrated into several parts of Greece. Some of the Pelasgians that had been driven from Attica, settled at Lemnos, whither, some time after, they carried some Athenian women, whom they had seized on the coast of Africa. They raised children by these captive females, but afterwards destroyed them together with their mothers, through jealousy, because they differed in manners from themselves, which horrid murder was attended by a dreadfnl pestilence. Such is the account given by the classic writers (Pausanias, Strabo, Herodotus, Plato, Virgil, Ovid, Flaccus, Seneca, &c.) But, when we weigh the foregoing arguments, can we doubt that the women were destroyed through jealousy of their religion, and not because they differed merely in manners, in accordance with the peculiar characteristics of fanaticism, which brooks no opposition to its devouring nature?

The word Pelargos was derived, says Wilford, from Phala, and Argha, (Phallus, and Argha from Bhaga, or Yoni), those mysterious types which the later mythologists distinguished under the names of Pallas and Argo.

The Pelargi venerated both male and female principles in union, as their compound appellation indicates, and represented them conjointly, when their powers were supposed to be combined, by the intersection of two equilateral triangles, thus, X, that peculiar symbol
the emblem of Lux, and to which innumerable perfections and virtues, including those of the Cross, have been attributed, from time immemorial. The union of these two symbols, devoting the male nature, and the female nature, or the phallus, the mark of which is the upright line, and the Yoni, the recognitive mark of which is the *horizontal* line, are best rendered, or depicted, in the double, or conjoined equilateral triangle in intersection. The pyramidal, aspiring, equilateral triangle is male, and signifies fire, and the rushing force of fire, mounting upwards in its own impulse, contradicting nature, inasmuch as it shoots up against gravity. The pyramid in reverse, or pointing down, is the indicating symbol of Wata, or of the lunar, female influence.

Sacti Puja, the worship of the female powers, presents us with a very singular phase of the extraordinary subject we are examining, and we now proceed to lay before our readers certain

"Form'd all mysteries to bear."

facts which will enable them to form some clear idea of its character. In England perhaps, save in learned circles, but little is known of this matter. India however has long been familiar with it, and, owing to certain revelations in the law courts in connection with an action for libel, has become notorious for the gravest scandals in connection therewith.

The creed Sacteya (Sharkt-ya) is said to be one of the most curious of the results of, or emanations from, the "austere principles inculcated by the Saiva and Vaishnava Codes of the Hiudu faith."

Sacti is power, and to this great goddess the Sectcya creed, though acknowledging Brahma, Vishnu, Siva, and others, declares all to be subject. The doctrines of this particular sect are to be found in the sacred books called Tautras, which until a comparatively late period were virtually unknown to most Europeans. By the aid of extracts from these books and the testimony of witnesses in the trial just alluded to, we shall be able to present a tolerably comprehensive view of this outrageously immoral and fanatical phase of a sexual worship which at one time was almost perfectly free from reproach.

This sect is known in India as the Maharajas, Rudra Sampradáya, or Pushti Marga. Vallabhácharya, the progenitor, was born about 400 years ago. Laxman Bhat, his father, was a Telinga Brahmin, from the country called Talingra, situated in Southern India. In a book of the Mahárájas called "Sampradáya-pradipa" (the illuminator of customs), it is stated that Luxman Bhat was promised by God, that he should have three sons, whereof the second would be his own (that is God's) incarnation. Accordingly after the birth of Rámkarshna, the first son, Laxman Bhat started on a pilgrimage to the Holy Places. Ere he reached Kásee (Benares), a serious quarrel had arisen between the Hindus and Mussulmans of that city; and Laxman Bhat, therefore, left that spot and arrived in a jungle called Champá. There his wife gave birth to a child on Sunday, the 11th of Vaisakh Vadya, Samvat 1535. As he and his family had to fly for their lives, the infant was left among the leaves of the trees, and they stopped in the town of Chandá. After a few weeks, when quietness was restored in Benares, they set out for that place by way of Champá Forest. When they came to the latter place, they saw a boy playing in the midst of a pit containing sacrificial fire, and taking that child to be their own, they took him along with them to Benares, and gave him the name of Vallabha. When he grew older and founded a new creed, he came to be known by the name of Vallabhácharya. A temple has been built on the spot where he was born.

From the time Vallabhácharya began to inculcate his own new creed called "Pushti Márga," up to the day he died, 84 Vaishnavas had agreed to follow his doctrines. Vallabhácharya had two sons, whereof Vithal Náthji is reckoned the principal incarnation of God. He brought 252 Vaishnavas into the new creed. Vithal Náthji had seven sons, who established themselves at seven different localities. They gave themselves out to be incarnations of God, and that Vallabhácharya came out playing from a pit of sacrificial fire, and by these means increased the number of their followers, who began to come over by thousands.

As the progeny of these seven increased, so the number of Mahárájas also was augmented. Soon as a child was born in the family of a Mahárál, he was called a Mahárál, and the Vaishnavas considered him to be an incarnation of the divinity, and fell at his feet. If a Mahárál died, the Vaishnavas could not say that he died, because he was a divine being; he was said, therefore, to have extended his career to the other world.

Amongst other articles of the new creed, Vallabha introduced one, which is rather singular for a Hindu religious innovator or reformer: he taught that privation formed no part of sanctity, and that it was the duty of the teachers and his disciples to worship their deity, not in nudity and hunger, but in costly apparel and choice food: not in solitude and mortification, but in the pleasures of society and the enjoyment of the world. The Gosains, or teachers, were almost always family men, as was the founder, for after he had taken off the restrictions of the monastic order to which he originally belonged, he married, by the particular order it is said, of his new god.

The Gosains were always clothed with the best raiment and fed with the daintiest viands by their followers, over whom they had unlimited influence; part of the connection between the Guru and teacher being the three-fold Samarpan, or consignment of Jan, Man, and Dhan, body, mind, and wealth, to the spiritual guide. The followers of the order were specially numerous amongst the mercantile community, and the Gosains themselves were often largely engaged, also, in maintaining connections among the commercial establishments of remote parts of the country, as they were constantly travelling over India, under pretence of pilgrimage, to the sacred shrines of the sect, and notoriously reconciled upon these occasions the profits of trade with the benefits of devotion; as religious travellers, however, this union of objects rendered them more respectable than the vagrants of any other sect.

The practices of the sect were of a similar character with those of other regular worshippers : their temples and houses had images of Gopál, of Krishna, and Rádhá, and other divine forms connected with this incarnation, of metal chiefly, and not unfrequently of gold, the image of Krishna represented a chubby boy, of the dark hue of which Vishnu is always represented : it was richly decorated and sedulously attended, receiving eight times a day the homage of the votaries. These occasions took place at fixed periods and for certain purposes, and at all other seasons and for any other object, except at stated and periodical festivals, the temples are closed and the deity invisible. The eight ceremonials are the following :—

1.—*Manglá*, the morning levee : the image, being washed and dressed, is taken from the couch, where it is supposed to have slept during the night, and placed upon a seat, about half-an-hour after sunrise ; slight refreshments are then presented to it, with betel and pan ; lamps are generally kept burning during this ceremony.

2.—*Sringára :* the image having been anointed and perfumed with oil, camphor and sandal, and splendidly attired, now holds his public courts : this takes place about an hour-and-a-half after the preceding, or when four Gheries of the day have elapsed.

3.—*Gwála :* the image is now visited, preparatory to his going out to attend the cattle along with the cow-herd ; this ceremony is held about forty-eight minutes after the last, or when six Gheries have passed.

4.—*Rája-Bhóga :* held at mid-day, when Krishna is supposed to come in from the pastures and dine ; all sorts of delicacies are placed before the image, and both these and other articles of food dressed by the ministers of the temple are distributed to the numerous votaries present, and not unfrequently sent to the dwellings of worshippers of some rank and consequence.

5.—*Uthápan :* the calling up ; the summoning of the god from his siesta ; this takes place at six Gheries, or between two and three hours before sunset.

6.—*Bhoga:* the afternoon meal, about half-an-hour after the preceding.

7.—*Sandhyá:* about sunset, the evening toilet of the image, when the ornaments of the day are taken off, and fresh unguent and perfume applied.

8.—*Sayan:* retiring to repose; the image, about eight or nine in the evening, is placed upon a bed, refreshments and water in proper vases, together with the betel box and its appurtenances, are left near it, when the votaries retire, and the temple is shut till the ensuing morning. *

Besides their public demonstrations of respect, pictures and images of Gopála are kept in the houses of the members of the sect, who, before they sit down to any of their meals, take care to offer a portion to the idol. Those of the disciples who have performed the triple Samar-pana, eat only from the hands of each other, and the wife or child that has not exhibited the same mark of devotion to the Guru can neither cook for such a disciple nor eat in his society.

The mark on the forehead consists of two red perpendicular lines, meeting in a semi-circle at the root of the nose, and having a round spot of red between them. The salutations amongst them are Krishna and Jaya Gopal.

Vallabha was succeeded by his son, Vitalla Nath, known amongst the sect by the appellation of *Sri Gosain Ji;* Vallabha's designation being *Sri Acharji.* Vitalla Náth, again, had seven sons—Girdhári Rai, Gouind Rai, Bál Krishna, Gokul Náth, Raghu Náth, Yadunáth, and Ghanssyama ; these were all teachers, and their followers, although in all essential points the same, form as many different communities. Those of Gokulnáth, indeed, are peculiarly separate from the rest, looking upon their own Gosains as the only legitimate teachers of the faith, and withholding all sort of reverence from the persons and Maths of the successors of his brethren : an exclusive preference that does not prevail amongst the other divisions of the faith, who do homage to all the descendants of all *Vitalla Nath's* sons.

The worshippers of this sect are very numerous and opulent, the merchants and bankers, especially those from Guzrat and Malwa, belonging to it. Their temples and establishments are numerous all over India, but particularly at *Mathura* and *Bindraban,* the latter of which alone is said to contain many hun-

* See H. H. Wilson's Works.

dreds, some of great opulence. In Benares were two temples of great repute and wealth, one sacred to *Lal Ji*, and the other to *Purushotama Ji*. Jaganath and *Dwarka* are also particularly venerated by this sect, but the most celebrated of all the Gosain establishments, was at *Sri Nath Dwár* in Ajmer. The image at this shrine is said to have transported itself thither from Mathura, when Aurangzeb ordered the temple it was there placed in to be destroyed. It is a matter of obligation with the members of this sect to visit *Sri Nath Dwár*, at least once in their lives; they receive there a certificate to that effect, issued by the head Gosain, and, in return, contribute according to their means to the enriching of the establishment. It is a curious feature in the notions of this sect, that all the veneration paid to their Gosains is paid solely to their descent, and unconnected with any idea of their sanctity or learning; they are not unfrequently destitute of all pretensions to individual respectability, but they not the less enjoy the homage of their followers.

The Maharajas have contrived various schemes to throw their followers in the trap and temptation of licentiousness. They have laid open to their followers the doings and examples of *Krishna*, and explained that their female worshippers will obtain salvation in the same way as the Gopis have obtained it by making *Rás-lila* (wanton or amorous sport with many women) with *Krishna*. They (the Maharajas) sit in their temples with the hairs of their heads fashionably dressed and in costly apparel, and rub over their bodies the essence of rose and aromatic fragrance. The female followers, who go to touch their feet, fall in temptation by their costly ornaments and odoriferous fragrance. They (the Maharajas) worship Thákurji (the image) for name's sake. This business is entrusted to their servants, and they merely go and stand near Thákurji, when everything is prepared. Instead of casting their look on Thákurji, they intently look at the female worshippers. They send sweetmeats and spiced milk to some of their sevkees (female devotees), and by such several contrivances they throw them in the trap of licentiousness.

The Maharajas have written in their books a great deal to support the cause of licentiousness. Their most orthodox followers meet together in the evening, to hear stories from books called the "tales of eighty-four and two hundred and fifty-two Vaishnavas." The males and females assemble to listen to these narratives. Some of the stories produce bad effects on the minds of the hearers, who are thus inclined to licentiousness. At last, so great is the effect that they become prepared to form a *Ras Mandali.*

This term has two meanings : the one is a society of love, affection, and fondness ; and the other is from *Ráslilá*. In *Ráslilá*, the Gopis used to mix, dance, and fall in love with *Krishna.* With this object the *Ras Mandali* is formed. These Mandalis (societies) do not meet publicly, but are convened privately at the residence of some orthodox and rich *Vaishnava*. In this Mandali, some of the *Vaishnavas* and their wives meet together and discuss matters of love and affection. They then bring before them the sweetmeats which have been consecrated to *Thákurji.* They then put a morsel in each other's mouth, and feed each other's wife.

The wife of one *Vaishnava*, puts a morsel in the mouth of another *Vaishnava*, who in turn does the same to her with ardent love. After they have done eating, they fall in so much love with each other that they join in promiscuous intercourse. One Dámodar Swámi of the Maharaja has written a verse regarding the Ras Mandali, which says :—" We must eat and drink with mutual love and take delight in sexual intercourse. This is a praise-worthy act of those, who have consigned their minds to Gokulnáthjee, and know that females are intended for men."

In this *Ras Mandali*, if a male *Vaishnava* wish to enjoy the wife of another *Vaishnava*, the latter should give that liberty with great delight and pleasure ; not the slightest hesitation is to be made. This is a chief condition with a *Vaishnava*, who wishes to be a member of this Mandali, and though he is without a wife, he can take liberties with the wives of other *Vaishnavas*. Captain MacMoode, resident in Kutch, has observed about the *Ras Mandalis :*—" The well-known *Ras Mandalis* are very frequent among them (Bhattias), as among other followers of Vishnoo. At these, persons of both sexes and all descriptions, high and low, meet together ; and under the name and sanction of religion practice every kind of licentiousness."

In this way the Maharajahs have spread a net of licentiousness among their followers.

"The word Tantram," says Mr. Edward Sellon, "signifies literally—art, system, craft, or contrivance ; prescribing the abolition of all caste, the use of wine, flesh, and fish (which the Brahminical code considers unlawful for Brahmins), with magical arts, diagrams, and the express adoration of the female sex. The Sacti sect is, in fact, what the Greeks called Telestica or Dynamica, and like gnosticism inculcates great contempt of the acknowledged religion, the peculiarities of which are only alluded to as matter for ridicule. Like gnosticism, it teaches magic, and looks upon the causes and agents of evil as the gods of the world. Let it not be supposed, however, that the creed of the *Sacteyas* is a religion of a modern date; the Brahmins look upon the books describing it as undoubtedly ancient,—more ancient, indeed, than the Purans. The most popular of these books are comprised in the following, to which are here given equivalent titles :—

1. *Sarada Tilacam.*—The Masterpiece.
2. *Jyan Arnavam.*—A System of Wisdom.
3. *Culanarvam.*—The Noble Craft of Thought.
4. *Gudha Culanarvam.*—The Hidden Part of the Noble Art.
5. *Bagala Tantram.*—The Litany of the *Vulva.*
6. *Ananda Tantram.*—The System of Joy.
7. *Rudra Yamalam.*—Conversations of Siva and his Spouse.
8. *Yogini Hriolayam.*—The Heart of the Angel—this is also called *Yoni Tantram.*
9. *Sir' Archana Chandrica.* — Rules for the worship of Girls in the bloom of youth.—[In the Calpam, cc. iii. and iv., is a description of every limb of a woman, with the *Madan a'layam*, and how they should be adored.
10. *Lyam' Archana Tarangini.*—The System of Worshipping a Girl.
11. *Anand Calpa Valli.*—The Rites of Delight.
12. *Tantra Saram.*—Summary of the Craft.
13. *Tantra Rajam.*—Illustrations of the Sublime Art.

"The system advocated in these books is termed Panchamacaram. In other words, the five mystical M's, in allusion to the five words beginning with M, viz.: Madya, Mamsa, Matsya, Mantra, Mithuna, *i.e.*, wine, flesh, fish, magic, and lewdness; which have reference to the following as a proposed means for the attainment of beatitude in the next world :—

1. A total freedom from cast and distinctions of every kind.
2. A liberty of eating flesh and fish, and drinking wine.
3. Promiscuous enjoyment.
4. The practice of magic, and the adoration of women.
5. The worship of demons and Yogini *i.e.*, Powers.

The *Sacteyas* are divided into two sects : the *Daxin' ácháram* (or right hand), and Vāmācharam (or left hand). Each sect renounces the established religion, and declares the worship of women supreme, every woman (according to them) being a Sacti, or image of the great goddess. There rules for fasting, bathing, and prayer, are to the full as irksome as with the Brahmins themselves. The person worshipped is a woman or girl of the Brahminical caste (among the *Daxin ácháram*), who is elegantly dressed, and adorned with jewels and garlands. One, three, or nine females are to be thus adored by one or more men; but in the left hand mode, there is only one girl and one worshipper.

"The *Vāmācharum* sect veil in deep mystery the rites which they practise. They commence by fasting and bathing like the *Daximácháruns;* but many of their observances are of a less innocent nature. The great feast, called *Siva Ratri*, is the period of the year when the Hindu worship of Venus is to be performed: other days are also named in their code besides the *Siva Ratri*, or *Dussera*. The person who wishes to perform the sacrifice is to select a beautiful young girl of any caste, a pariah, a slave, a courtesan, or nautch girl, would be preferred. She is called *Duti*, or 'angel messenger,' or conciliatrix, being the medium of intercourse between the worshipper and the goddess. She is also called *Yogini*, or nun,—literally, 'one who is joined.' The Yogini Hridayam, or 'Heart of the Nun,' is a book well known to these sectaries; it is usually known by the name *Yoni Tantrum*, or, 'Ritual of Vulva Worship,' *Yogini* being used as an occult name of Yoni (pudendum muliebre). It is a peculiarity that no widow, however young and lovely, is ever selected. After fasting and bathing, she is elegantly dressed and seated on a carpet. The fine acts—already mentioned in alluding to the letter M—are then performed in order, and the votary erects a magical diagram and repeats a spell. These diagrams are diverse. The spell called *Agni Puram* has for a diagram a 'volcano,' *i.e.*, a double circle, and therein a triangle, doubtless the same with the *atish kadr*, or 'house of fire.' Spells are always used. The devotee next meditates on her as Pracriti (Nature), and on himself as a deity. He offers prayer to her, and then proceeds to inspire her in each particular limb with some one goddess, of the host of goddesses. He adores, in imagination, every individual part of her person, and, by incantation, lodges a fairy in every limb and member, and one in the Yoni, as the centre of delight. The names of the female sylphs addressed to her are not very delicate, and need not be here further alluded to. Then follow the second, third, and fourth M, *i.e.*, he presents her with flesh, fish, and wine. He makes her eat and drink of each, and what she leaves, he eats and drinks himself. He now divests her entirely of all clothing, and himself also. He recommences to adore her body anew in every limb; from this the rite is often termed *Chacra Puja*, or worship of the members. He finally adores the Agni Mandalam with reverent language, but lewd gesticulations. Special rites are used, says the Ananda Tantra, to divest her of all shame, and shame can only be annihilated by the use of wine."

In the Sri Vidya, which enjoins secrecy, occurs the following :—

"Such was the rule, sung by the inspired prophets. For those who adore the young and lovely Sacti, revealed to none but the few. Keep it concealed, like the rosy lips that pout between the recess of thy thighs, O Goddess. Hide this creed, so pure so excellent, as closely as you hide your vulva cleft. O hide this code of bliss, lady, from vulgar eyes."

Again, in the *Agni Mandalam* [volcano].

"Let the fuel of sacrifice be her decorations; let the altar of sacrifice be her middle; the pit of her navel is the hearth, and her mouth the ceaseless fire; the south point her chin; her rosy hand the spoon; let the Sabhya and Avasadlya be the two sides of the same. The holy flame is the moist vulva. The fuel is collision (because fire is produced by friction), and the Lord Linga is the great high priest."

Again in the Cama Cala,—

1. "Let us land the god and goddess *Racta* (Parvati) and *Sucla* (Siva), ever glorious! Primary, noblest of fanciful blisses, without compare! highest in glory, which is comprehended by the wise alone.

2. "To the great and holy one, accomplished in voluptuous movements, elevated in enjoying! The *tejas* compounded of blood and ——, to him I bow! Praise him, the supreme Lord of delight! noblest in faith, the only bliss of my soul *(Madia pamam)*, the most secret *Vedha*, veneration.

4. "The bliss of all men, exalted on his throne. To Siva, my Lord, soul viewed, the form of bliss, the glorious! may he with his slant glance, remove the foulness of mistrust. By the holy act of enjoyment was the blessed science called amorous,

aroma invented. How can it be denominated? The unmentioned," &c.

[Then begins the book called the Spirit of Sexual Joy].

Mr. Sellon then supplies us with some further extracts from these books from which we call the following. Passing over certain introductory matters we proceed thus:—

3. "Unspeakable, incomparable in form, inexpressible by writing, by figure, or by image.

4. "That sun, the supreme Siva *(i.e., Sucla)*, whose rays are reflected in the heart, that in reflecting the glorious beauty, receives the great seed.

5. "That sense of individuality which is inherent in the mind, clearly expressed in the term A' HARNAM [A denotes *Siva, i.e.*, semen; and HA denotes *Sacti, i.e.*, Power, typified by blood, the two are united by the mystic word.]

6. "Whiteness and redness (blood) when their respective fluids are united, a word and its import; so are united creation and its cause, mutually collocated and indivisible.

7. "The fluid is the source of individuality, and the (portion) abode of the sun is therein; and *Cama* (or Cupid) being the attractive power, is the *Cula* (spirit), and is the enjoyment.

8. "This is the discrimination of *Cula* (male and female joined *in coitu*), and is equivalent with *Sri Chacra*. He who knows to distinguish them is the freed, and shall assume the form of the great *Tripuri*.

9. "There is distilled from the red *Sacti* the mystical sound CLIM, which is denominated *Nada Brahma*, and the sound is audible; from it originates the ether, wind, and fire, and the terrestrial decade.

10. "Next, from the fluid thus made known, spring wind, fire, water, and earth, all the universe, from an atom up to a sphere.

* * * * * * * * * *

13. "The three great powers are those of *Desire*, of *Knowledge*, of *Loving*, and of performing the act.

14. "And in the same order are three Lingas, of tangible (Sthula) spiritual bodies, and visionary, and this is *Tripura*, triple; and the forth is the Art.

15 "Sound, touch, form, taste, and smell, and the essential qualities of each multiplied by the three *gunas* (qualities) of *Prithoi*.

16. "Hence the *Spell* of fifteen syllables.
17. "And there are fifteen *Tithis*.
18. "On the letters, consonants, and vowels.
19. "The Art is magic; the object is the goddess.
20. "From letter Y to letter S there are three forms.
21. Between the *chacra* (members) and the goddess it is impossible to draw any distinction before the spiritual body is evolved.
22. "In the centre of the *chacra* let the mystic fluid be; this is the essential fluid.
23. "The three that are formed from the triple root," &c.

From the passages here cited, it will be seen how closely the Sacteya rites resemble those practised by ancient Pagan peoples; they are expressly forbidden in the Mosaic law. "Ye shall not eat anything with the blood, neither shall ye use enchantment, nor observe times"—Lev. xix. 26. "Giving his seed unto Moloch;" "Who go a-whoring after idols;" "Turning after familiar spirits, to go a-whoring after them"—Lev. xx. 2, 6.

Mr. Sellon further says:—"The diagram, also, discovered by Cicero, on the tomb of Archimedes, appears identical with one of those spells used in the *Sacti Puja*,—the apex of the triangle is downwards, with a point in the centre.

"In India, the adorers of the goddess regard the mystical ring, or circle, as the orifice of the vagina, while the triangle represents the nymphæ, the dot represents the fairy lodged in this member. When the imagination of the *Sacti* is sufficiently excited by wine, divine homage, and libidinous excess, she is supposed to be in a *guyana nidra*, or mystic sleep, wherein, like the sibyls among the ancients and the modern clairvoyants, she answers questions in a delirious manner, and is supposed to be, for the time, the mouthpiece of the deity.

"Such is the *Sacti Puja*, or worship of power, power here meaning the good goddess Maya (delusion); she is also called *Bagala, Vagala*, and *Bagala Mukhi*. She has neither images nor pictures, and is usually typified by a vessel of water. The girl who performs *Sacti* (for the time) is the only true representative of the goddess.

"The Eleusinian mysteries bear a very striking analogy to the Sacteya, and those writers err who have asserted that the mysteries of Eleusis were confined to men. A reference to d'Hancarville (Naples edit., 1765, tome IV.) will give several instances of the initiation of women. The method of purification, portrayed on antique Greek vases, closely resembles the ceremony as prescribed in the Sacti Sodhana. From this circumstance, and also from the very frequent allusions to *Sacteya* rites in the writings of the Jews and other ancient authors, it is evident that we have now in India the remains of a very ancient superstitious mysticism, if not one of the most ancient forms of idolatry, in the *Sacti*, or *Chacra Puja*, or worship of Power."

These particulars, so far as Mr. Sellon's name is connected with them, were communicated by a very learned orientalist, a member of the Madras Civil Service, and a judge, whose name it is impossible to give owing to a condition he himself imposed.

The worshippers of the Sakti, the power or energy of the divine nature in action, are exceedingly numerous among all classes of Hindus. This active energy is, agreeably to the spirit of the mythological system, personified, and the form with which it is invested, considered as the special object of veneration, depends upon the bias entertained by the individuals towards the adoration of Vishnu or Siva. In the former case the personified Sakti is termed Lakshmi, or Maha Lakshmi, and in the latter Parvati Bhavam, or Durga. Even Sarasvati enjoys some portion of homage, much more than her lord Brahma, whilst a vast variety of inferior beings of malevolent character and formidable aspect receive the worship of the multitude. The bride of Siva, however, in one or other of her many and varied forms, is by far the most popular emblem in Bengal and along the Ganges.

The worship of the female principle, as distinct from the divinity, appears to have originated in the literal interpretation of the metaphorical language of the Vedas, in which the will or purpose to create the universe is represented as originating from the creator, and co-existent with him as his bride, and part of himself. Thus, in the Rig Veda it is said, "That divine spirit breathed without afflation, single with (Svadhá) her who is sustained within him; other than him nothing existed. First, desire was formed in his mind, and that became the original productive seed;" and the *Sama Veda*, speaking of the divine cause of creation, says, "He felt not delight, being alone. He wished another, and instantly became such. He caused his own self to fall in twain, and thus became husband and wife. He approached her, and thus were human beings produced." In these passages it is not unlikely that reference is made to the primitive tradition of the origin of mankind, but there is also a figurative representation of the first indication of *wish* or *will* in the Supreme Being. Being devoid of all qualities whatever, he was alone until he permitted the wish to be multiplied, to be generated within himself. This wish being put into action, it is said, became united with its parent, and then created beings were produced. Thus this first manifestation of divine power is termed *Ichchharupa*, personified desire, and the creator is designated as Svechchhámaya, united with his own will; whilst in the *Vedánta* philosophy, and the popular sects, such as that of Kabir and others, in which all created things are held to be illusory, the Sakti, or active will of the deity, is always designated and spoken of as Máyá, or Mahámáyá, original deceit or illusion.

Another set of notions of some antiquity which contributed to form the character of the Sakti, whether general or particular, were derived from the *Sankhya* philosophy. In this system, nature, Prakriti, or Múla Prakriti, is defined to be of eternal existence and independent origin, distinct from the supreme spirit, productive, though no production, and the plastic origin of all things, including even the gods. Hence Prakriti has come to be regarded as the mother of gods and men, whilst as one with matter, the source of error, it is again identified with Máyá, or delusion, and as co-existent with the supreme as his *Sakti*, his personified energy or his bride.

These mythological fancies have been principally disseminated by the *Puranas*, in all which *Prakriti*, or Máyá, bears a prominent part. The aggregate of the whole is given in the *Brahma Vawartta Purana*, one section of which, the *Prakriti Khanda*, is devoted to the subject, and in which the legends relating to the principal modifications of the female principle are narrated.

According to this authority, Brahma, or the supreme being, having determined to create the universe by his super-human power, became two-fold, the right half becoming a male, and the left half a female, which was *Prakriti*. She was of one nature with Brahma. She was illusion, eternal and without end: as is the soul, so is its active energy; as the faculty of burning is in fire. In another passage it is said, that Krishna, who is in this work identified with the Supreme, being alone invested with the divine nature, beheld all one universal blank, and contemplating creation with his mental vision, he began to create all things by his own will, being united by his will, which became manifest, as

Mula Prakriti. The original Prakriti first assumed five forms. Durga the bride, *Sakti*, and *Máyá*, of Siva, Lakshmi the bride, Sacti and Máyá of Vishnu, Sarawasti the same of Brahma, or in the *Brahma Vaivartta Puránа*, of Hari, while the next Savitri is the bride of Brahma. The fifth division of the original Prakriti, was Radha, the favourite of the youthful Krishna, and unquestionably a modern intruder into the Hindu Pantheon.

Besides these more important manifestations of the female principle, the whole body of goddesses and nymphs of every order are said to have sprung from the same source, and indeed every creature, whether human or brutal, of the female sex, is referred to the same principle, whilst the origin of males is ascribed to the primitive *Purusha*, or male. In every creation of the universe it it is said the Mula Prakriti assumes the different gradations of *Ansarúpini*, Kalarupini, Kalánsarupini, or manifests herself in portions, parts, and portions of parts, and further subdivisions. The chief *Ansas* are, besides the five already enumerated, Ganga, Tulasi, Manasá, Shashthi- or Davasená, Mangalachandika, and Kali ; the principal *Kalas* are Swaha, Swadha, Dakshiná, Swasti, Pushti, Tushti, and others, most of which are allegorical personifications, as *Dhriti*, Fortitude, *Pratishtha*, Fame, and *Adharma*, Wickedness, the bride of *Myrityn*, or Death. Aditi, the mother of the Gods, and Diti the mother of the Demons, are also *Kalas* of Prakriti. The list includes all the secondary goddesses. The *Kalansas* and *Ansánsas*, or subdivisions of the more important manifestations, are all womankind, who are distinguished as good, middling, or bad, according as they derive their being from the parts of their great original in which the *Satya Rajas*, and *Tama Guna* or property of goodness, passion, and vice predominates. At the same time as manifestations of the great cause of all they are entitled to respect, and even to veneration : whoever, says the *Brahma Vaivartta Puránа*, offends or insults a female, incurs the wrath of Prakriti, while he who propitiates a female, particularly the youthful daughter of a Brahman, with clothes, ornaments, and perfumes, offers worship to Prakitri herself. It is in the spirit of this last doctrine that one of the principal rites of the *Saktas* is the actual worship of the daughter or wife of a *Brahman*, and leads with one branch of the sect at least to the introduction of gross impurities. But besides this derivation of Prakriti, or Sakti, from the Supreme, and the secondary origin of all female nature from her, those who adopt her as their especial divinity employ the language invariably addressed towards the preferential object of worship in every sect, and contemplate her as comprising all existence in her essence. Thus she is not only declared to be one with the male deity, of whose energy some one of her manifestations is the type, as Deni with Siva, and Lakshmi with Vishnu; but it is said, that she is equally in all things, and that all things are in her, and that besides her there is nothing. *

The profligacy, debauchery, and licentiousness which characterize the sect of Vallabhácháryá have been noticed by several distinguished persons, two or three of whom flourished some hundred years ago. Dámodar Svámi, a dramatic writer, composed a Sanskrit drama entitled *Pakhanda Dharma Khandan*, in the year Samvat 1695 (about A.D. 1639), in which a distinct reference to Vallabhácháryá and his sect is made as follows:—

"The Sútradhára (says to the Nati):—O dear, the Vedas have fled somewhere; no one knows the story of their flight (*i.e.*, whither they have gone). The collection of the Sankhya, Yoga, and the Puranas, has sunk into the bowels of the earth. Now, young damsels, look to the self-dedication preached by Shrimat Vallabha Vittaleshvara, who has conspired to falsify the meaning of the Vedas.

Enters a Vaishnava, having on his neck, ear, hand, head, and around his loins, a wreath made of the *Vrinda*, having on his forehead *Gopichandana* (a substitute for sandal-wood). He is the one who repeats Rádhá! Krishna! Being opposed to the Shruti, he is the reproacher of those who adhere to the Vedas. He finds at every step crowds of females filled by *Káma* (lust or cupid). He is the Kisser of female Vaishnavas. Ye Vaishnavas, ye Vaishnavas, hear the excellent and blessed Vaishnava doctrine: the embracing and clasping with the arms the large-eyed damsels, good drinking and eating, making no distinction between your own and anothers, offering one's self and life to gurus, is in the world the cause of salvation.

Mutual dining, intercourse with females night and day, drinking, forming endless alliances, are the surpassing, beautiful customs of the persons who have consecrated their souls to Sri Gokulesha. Charity, devotion, meditation, abstraction, the Vedas, and a crore of sacrifices are nothing: the nectarine pleasure of the worshippers of the *Páduká* (wooden slipper), in

* See H. H. Wilson's Works, vol. I.

Sri Gokula, is better than a thousand other expedients. Our own body is the source of enjoyment, the object of worship reckoned by all men fit to be served. If intercourse do not take place with the Gokulesha, the paramour of men is useless, like a worm or ashes.

The chief religion of the worshippers of the Paduka is the consecration of a daughter, a son's wife, and a wife, and not the worship of Brahmanas learned in the Vedas, hospitality, the Shraddha, vows and fastings."

The following is a specimen of several stories written in a religious book of the Maharajahs, known as "Chorási Vaishnava Ki Bártá (the Stories of Eighty-four Vaishnavas), It is intended to show that a devotee is bound at any rate, and if female, even at the sacrifice of her chastity, to become serviceable not only to Acháryaji (Maharaja), but also to the time Vaishnavas or believers in Maharajas :—

"Krashnadás (a follower of Vallabha) lived in a town. He was a pious and reserved man. The followers of Shri Achariaji Maha Prabhú (Vallabha) lived in various villages. They used to go in company to pay *darshna* (divine homage), to Mahá Prabhú to a village called Adel. At one time, about ten or twenty Vaishnavas meeting together, set out to pay homage to Maha Prabhú. They arrived at the village in which Krashnadás lived. When they entered the house of Krashnadás, he was not there. He had gone on business to a village situated at the distance of two or three *Kos*. The wife of Krashnadás was in the house. She bowed to her visitors, and after making the usual salutation of Shri Krishna, they were requested with great reverence to take their seat. Afterwards, going in her chamber, she began to think for herself as to what she was to do. At that time she recollected that a certain Vániá shop-keeper was enamoured of her and he used to say if I cohabited with him he should give me whatever I asked for. 'To-day I will go to his shop and get some articles of food and say to him to give me the things I want to-day for food.' So thinking, the female walked on and arrived at the shop. The Vániá (shop-keeper) made some becks or signs of love, upon which the female said to the Vániá, 'I shall once meet you (in the embrace of love), give me to-day what I want for food.' The Vániá replied 'I would believe if you give me a promise.' Then the female gave a promise, and took whatever articles of food she wanted. Afterwards arriving at home, having cooked the food and dedicated it to Thákurji (the image), she fed the Vaishnavas. Then the Vaishnavas ate the feast with great joy. Afterwards Krashnadás came home in the evening. He paid his respects to all the Vaishnavas in person. He went in his chamber after making the usual salutation of Shri Krishna. He asked his wife as to whether she had feasted the guests. The female replied she had done so; and related to him all the matters that had passed. Then Krashnadás greatly became pleased with his wife. And then the wife and the husband feasted together. Afterwards Krashnadás came and sat in the assembly of the Vaishnavas. The whole night they talked about God. The day then broke; the Vaishnavas went away after taking leave of Krashnadás. Krashnadás saw them in person to a certain distance. Returning home, and after making his ablutions and worshipping Thákurji (the image), he went out (on business). The female after cooking the food, dedicated it to Thákurji, and kept the same under a cover. Krashnadás came home in the evening, when he and his wife feasted together. Then Krashnadás spoke thus to his wife:—'The Vaishnava to whom you have given a promise must be waiting for you. You must fulfil your promise and that is proper.' The female having bathed and applied *anjan* (black powdery substance) to the lower part of her eyes (which is the usual fashion in India of making them more attractive), and after having dressed herself most gaudily, with all proper adornments, she set out. It was the rainy season and it had ceased raining then. In consequence of this the roads were dirty and muddy. Therefore Krashnadás said to his wife, 'I shall carry you on my shoulder; otherwise you shall dirty your feet and therefore the Vániá would not like you.' So placing his wife on his shoulders, he took her to the shop of the Vániá, and there alighted her. Afterwards the female called aloud the Vániá, and asked him to open the doors. Opening the doors, the Vániá took her in. He brought water and desired to wash her feet. The female then told 'my feet are not dirty.' On this the Vániá inquired 'how it was that while the road was dirty, your feet are clean.' The female then said to the Vániá, 'what business have you with that question or matter; do your business (meaning that for which the meeting was appointed.') The Vániá then said the whole truth must be related to him. The female then related that her husband brought her there on his shoulders. The Vániá was wonder-struck when he heard the whole story. He took the female to her husband and implored forgiveness of him. He then became a follower of Maha Prabhú."

A few extracts from the opinions of the press, published some years ago, will throw light upon the character of this notorious sect :—

"The Gosainji Maharajas of the Vaishnavas of this place, instead of giving religious instructions, carry on debaucherous practices on their followers. This appears nothing, looking on them with the eyes of a savage man ; but, thinking justly, it appears a wicked practice. These Maharajas appear totally divine to the Hindus, but their acts seem extremely base, and their heart full of sin, and their conduct out of the way of social arrangements, and their practices opposed to religion. Their followers expose the vices of their religious guides with respect to all this. . . . . Oh, Shiva ! Shiva ! that aged matrons like their (Maharajas) mothers, young women like their sisters, and maidens like their daughters, who come to touch the feet of these true religious guides in their temples, who come to pay *darshana* (divine homage) to these godly Maharajas, who repair to pay *darshana* believing them to be God, that they should be made victims of carnal intercourse by the Maharajas, instead of giving them religious instruction. Fie ! Fie ! upon this incarnation, oh ! damned (burnt) your Vaishnava religion."

Bombay Chabuk: June, 1859.

"Not only are their bodies and wealth dedicated to the service of these Maharajas, but their daughters, sisters, and wives, with their persons, are dedicated to these debaucherous religious preceptors. The authority of the Maharajas is exercised over their followers without any restraint."

The Prabhodaya: August, 1859.

"I have seen the deceit of the Maharajas ; now, lady ! none of you should go into the Maharajas' temple. Inviting a girl of tender age, they give the sacred sweetmeats, and representing the story of the Káhn Gopies, make a wanton assault. If they see wealth, they invite with affection, otherwise they heed not ; robbing the wealth and bloom of youthful beauty ! See the honesty of these religious instructors."

"The conduct of the Maharajas of the present day is so notorious that it is not necessary to say much about it. Besides, their acts are so disgraceful that our pen does not move to describe them in this work. Being possessed of affluence, they are, from their childhood, brought up in indulgence, and are allowed to do as they fancy, and receive no education whatever; most of the present Gosainjis (Maharajas), therefore, are ignorant fools; they do not possess as much knowledge as is required for the office of a guru. What admonition can one impart to others who does not himself possess any knowledge. . . . . The Gosainjis pass their time in eating daintiest viands, in wearing fancy clothes and jewels, in driving carriages, in committing adultery with strange women, and in repose."

Ancient Religion of the Hindus, 1861.

Captain MacMurdo, the Resident in Kutch says:—"The principal Maharaja at present, on this side of India, is named Gopinathji, a man worn to a skeleton, and shaking like a leaf from debauchery of every kind except spirituous liquors. He is constantly in a state of intoxication, from opium and various other stimulants which the ingenuity of the sensual has discovered."

Transactions of Literary Soc., Bombay, Vol. II.

"The Maharajas, for these evil purposes, through certain females and males, order sooner or later the female whom they have singled out from those who have come to pay *darshana* (divine homage). Sumptuously-dressed females, who are wantons, are invited by the Maharajas merely with a beck of their eyes. An invitation from the Maharaja is an invitation from Krishna, and thinking she has met God, she hastens with delight and precipitation to touch the person of the Maharaja. . . . . In these purposes, they (the Maharajas) do not use females of their own age, but upon tender youthful girls they exert their beastly strength."

"At this time a few Maharajas may be going in the right path. The majority of them follow the wrong path. The youthful fops are given to ostentation. The present children of Vallabha disgrace the name of their ancestors. . . . . The lalji (showy) Maharajas, when the darshana time has commenced and people crowded, sit in their bed-chamber inside the temple, and by the gesture of the eyes, or through some persons kept for the purpose, invite the female designed (for evil purposes), and commit evil act with her. In Surat once, a Maharaja, exerting his wild strength upon a girl who had not attained the age of puberty, had almost caused her death. Similar horrible events have happened at (Kutch) Mandvee, with which the Raja and his subjects of that country are not unacquainted."

The Guru and Woman, 1858.

"The temple of the Hindu Marajas is proved a brothel; their private dwelling, the home of a corrupt and disrespectable family; their eyes, wanton licentiousness ; their senses, the seat of wicked appetites (desires) ; every pore of their body, unrighteousness, uncleanliness, dirtiness ; and, in short, they have been found incarnations of devils, and possessed of the qualities of Satan instead of the incarnation of God."

The Apektyar : June, 1859.

"These Maharajas, claiming to be your spiritual guides, enjoy your young daughters and sisters, destroy your domestic comfort, and stain your character. Therefore, you Vaishnavas should keep anxiety about it, and, as the reformed party of your caste have used their prudence to shun these refuges, it behoves you to be on your guard. It is a credit to you to keep off your females from these debaucherous (Maharajas), and to observe the dictates of religion with prudence."

The Parsi Reformer: May, 1861.

"Most of the simple and ignorant female devotees are entrapped into this religious snare, and, giving money to Maharajas, practise adultery with them. But those immoral creatures, the Maharajas, are not content with this, and they many a time use violence on the tender body of the maidens (of their devotees), the instances of which are not uncommon. Such are these Maharajas—the pretended preceptors of religion, and their acts."

The Khoja Dost : August, 1861.

THE END.

www.ingramcontent.com/pod-product-compliance
Lightning Source LLC
Chambersburg PA
CBHW030406170426
43202CB00010B/1513